ADVANCED MATHS F[

Further Pure

Brian Gaulter and Mark Gaulter

Course consultant: Tony Dearman

FP1

OXFORD
UNIVERSITY PRESS

OXFORD

UNIVERSITY PRESS

Great Clarendon Street, Oxford OX2 6DP

Oxford University Press is a department of the University of Oxford.
It furthers the University's objective of excellence in research, scholarship,
and education by publishing worldwide in

Oxford   New York

Auckland   Cape Town   Dar es Salaam   Hong Kong   Karachi
Kuala Lumpur   Madrid   Melbourne   Mexico City   Nairobi
New Delhi   Shanghai   Taipei   Toronto

With offices in

Argentina   Austria   Brazil   Chile   Czech Republic   France   Greece
Guatemala   Hungary   Italy   Japan   Poland   Portugal   Singapore
South Korea   Switzerland   Thailand   Turkey   Ukraine   Vietnam

Oxford is a registered trade mark of Oxford University Press
in the UK and in certain other countries

British Library Cataloguing in Publication Data

Data available

ISBN: 978 0 19 914985 8

10   9   8   7   6   5   4

Typeset by Tech-Set Ltd, Gateshead, Tyne and Wear
Printed and bound in Great Britain by Bell and Bain.

Acknowledgements
The publishers would like to thank AQA for their kind permission to reproduce
past paper questions. AQA accept no responsibility for the answers to the past
paper questions which are the sole responsibility of the publishers.

The publishers would also like to thank James Nicholson for his authoritative
guidance in preparing this book.

The photograph on the cover is reproduced courtesy of Photodisc.

About this book

This Advanced level book is designed to help you get your best possible grade in the AQA MFP1 module for first examination in 2005. This module can contribute to awards in Pure Mathematics and Further Mathematics, at both AS level and A level GCE.

Each chapter starts with an overview of what you are going to learn and a list of what you should already know, with links where appropriate to Core modules. The 'Before you start' section contains 'Check in' questions, which will help to prepare you for the topics in the chapter.

You should know how to ...	Check in
1 Find the gradient of a line joining two given points.	**1** Find the gradient of the line AB when: a) A is $(1, 3)$ and B is $(3, 7)$ b) A is $(-1, 3)$ and B is $(-2, 2)$ c) A is $(0, 5)$ and B is $(5, 5)$

Key information is highlighted in the text so you can see the facts you need to learn.

> In general, if z is a complex number, its complex conjugate is denoted by z^*.
>
> If $z = x + iy$, then $z^* = x - iy$.

Worked examples showing the key skills and techniques you need to develop are shown in boxes. Also hint boxes show tips and reminders you may find useful.

Example 2

Find the general solution, in degrees, of the equation $\cos 5\theta = \dfrac{\sqrt{3}}{2}$.

$\dfrac{\sqrt{3}}{2}$ is the cosine of $30°$.

So, the general solution for 5θ is:

$$5\theta = 360\,n° \pm 30°$$

Dividing by 5, the general solution for θ is:

$$\theta = 72\,n° \pm 6°$$

Remember to find the **general solution for 5θ first**. Then transform the equation to give the general solution for θ.

You can check these values on a graphics calculator, after selecting the correct **range** or **view window**.

The questions are carefully graded, with lots of basic practice provided at the beginning of each exercise.

Towards the end of an exercise, you will sometimes find underlined question numbers. These are optional questions that go beyond the requirements of the specification and are provided as a challenge.

At the end of each chapter there is a summary. The 'You should now know' section is useful as a quick revision guide, and each 'Check out' question identifies important techniques that you should remember.

You should know how to ...	Check out
1 Find the sum and product of the roots of a quadratic equation.	**1** Find the sum and product of the roots of each of the following quadratic equations: a) $x^2 - 6x + 8 = 0$ b) $2x^2 + 5x - 1 = 0$

Following the summary you will find a revision exercise with past paper questions from AQA. These will enable you to become familiar with the style of questions you will see in the exam.

The book also contains a Practice Paper. This will directly help you to prepare for your exams.

At the end of the book you will find numerical answers and an appendix on complex numbers.

Contents

1 Roots and coefficients of a quadratic equation

This chapter will show you how to

✦ Find the sum and product of the roots of a quadratic equation
✦ Evaluate symmetrical functions of these roots
✦ Form new equations with roots related to those of the original equation

Before you start

You should know how to ...	Check in
1 Solve quadratic equations.	**1** Solve the following quadratic equations: a) $x^2 + 4x + 3 = 0$ b) $x^2 + 4x + 4 = 0$ c) $2x^2 - 3x + 1 = 0$ d) $3x^2 = 4x - 1$
2 Expand the square of a binomial expression.	**2** Expand and simplify: a) $(x + 1)^2$ b) $(2x - 3)^2$ c) $\left(x + \dfrac{1}{x}\right)^2$
3 Manipulate algebraic fractions.	**3** Write each of the following as a single algebraic fraction: a) $\dfrac{1}{a} + \dfrac{1}{b}$ b) $\dfrac{1}{a} - \dfrac{1}{b}$ c) $\dfrac{1}{a} \times \dfrac{1}{b}$ d) $\dfrac{a}{b} + \dfrac{b}{a}$ e) $\dfrac{a}{b} \times \dfrac{b}{c}$ f) $\dfrac{a}{bc} - \dfrac{b}{ac}$

Links to Core modules

The use of the discriminant and the factor theorem are in module C1. Knowledge of binomial expansions from module C2 would be helpful but is not assumed. Note that section 1.3 can be omitted as it will not be examined.

1.1 Roots of a quadratic equation

You solve a quadratic equation $ax^2 + bx + c = 0$ by finding its roots.

If α and β are the roots, then the expression $ax^2 + bx + c$ must have both $x - \alpha$ and $x - \beta$ as factors.

Thus $ax^2 + bx + c = 0$ is identical to $k(x - \alpha)(x - \beta) = 0$ for some constant k.

Therefore,

$$k(x - \alpha)(x - \beta) = ax^2 + bx + c$$

Expanding and collecting like terms gives:

$$k(x^2 - [\alpha + \beta]x + \alpha\beta) = ax^2 + bx + c$$

Equating the coefficients of x^2 gives: $k = a$

Equating the coefficients of x gives: $-k(\alpha + \beta) = b$

And equating the constants gives: $k\alpha\beta = c$

This leads to an important result.

> **Remember**
>
> The Factor Theorem
> If $f(\alpha) = 0$ for a polynomial $f(x)$,
> then $(x - \alpha)$ is a factor of $f(x)$.

FP1

For a quadratic equation $ax^2 + bx + c = 0$ with roots α and β:

$$\alpha + \beta = -\frac{b}{a} \quad \text{and} \quad \alpha\beta = \frac{c}{a}$$

Or

The **sum** of the roots of a quadratic equation is $-\dfrac{b}{a}$ and

the **product** of the roots is $\dfrac{c}{a}$.

Example 1

In the equation $3x^2 - 6x - 2 = 0$, find

a) the sum of the roots

b) the product of the roots.

....

a) Using $\alpha + \beta = -\dfrac{b}{a}$,

   Sum of the roots, $\alpha + \beta = -\dfrac{-6}{3} = 2$

b) Using $\alpha\beta = \dfrac{c}{a}$,

   Product of the roots, $\alpha\beta = -\dfrac{2}{3}$

You can check these results by evaluating the roots of the quadratic equation:

$$3x^2 - 6x - 2 = 0$$

and then finding their sum and product.

Using the formula for the solution of a quadratic equation:

$$x = \frac{6 \pm \sqrt{36 + 24}}{6} = \frac{6 \pm \sqrt{60}}{6} = 1 \pm \frac{\sqrt{15}}{3}$$

Hence, you can now determine α, β, $\alpha + \beta$ and $\alpha\beta$, as follows:

$$\alpha = 1 + \frac{\sqrt{15}}{3} \quad \text{and} \quad \beta = 1 - \frac{\sqrt{15}}{3}$$

$$\alpha + \beta = 1 + \frac{\sqrt{15}}{3} + 1 - \frac{\sqrt{15}}{3}$$

$$= 2 \quad \text{as expected.}$$

$$\alpha\beta = \left(1 + \frac{\sqrt{15}}{3}\right)\left(1 - \frac{\sqrt{15}}{3}\right)$$

$$= 1 - \frac{15}{9} = -\frac{6}{9}$$

$$= -\frac{2}{3}, \text{ as expected.}$$

> **Remember**
> For a quadratic equation
> $ax^2 + bx + c = 0$, you can use
> the formula:
> $$x = \frac{-b \pm \sqrt{b^2 - 4ac}}{2a}$$

> $\sqrt{15}$ is a **surd** (an irrational root).
> For the purposes of accuracy, it
> is best left in this form rather
> than approximated as a decimal.

FP1

Consider again the equation $ax^2 + bx + c = 0$. Divide through by a:

$$x^2 + \frac{b}{a}x + \frac{c}{a} = 0$$

This leads to a general result:

> $x^2 - \textbf{(sum of roots)}x + \textbf{(product of roots)} = 0$

You can use this result to derive a quadratic equation if you are given
information about its roots.

Example 2

Find the quadratic equation whose roots have a sum of $\frac{1}{2}$ and a
product of $-\frac{5}{2}$.

....

Using $x^2 - \text{(sum of roots)}x + \text{(product of roots)} = 0$,

$$x^2 - \frac{1}{2}x - \frac{5}{2} = 0 \quad \text{or} \quad 2x^2 - x - 5 = 0$$

Example 3

The quadratic equation $3x^2 + 9x - 11 = 0$ has roots α and β.
Find the equation whose roots are $\alpha + \beta$ and $\alpha\beta$.

....

From $3x^2 + 9x - 11 = 0$,

$$\alpha + \beta = -3 \quad \text{and} \quad \alpha\beta = -\frac{11}{3}$$

The sum of the **new roots** is: $\alpha + \beta + \alpha\beta = -3 - \dfrac{11}{3} = -\dfrac{20}{3}$

The product of the **new roots** is: $(\alpha + \beta) \times \alpha\beta = -3 \times -\dfrac{11}{3} = 11$

Therefore, the new equation is:

$$x^2 + \frac{20}{3}x + 11 = 0 \quad \text{or} \quad 3x^2 + 20x + 33 = 0$$

Exercise 1A

FP1

1 Write down the sum and the product of the roots of each of these quadratic equations.

   a) $x^2 + 3x - 7 = 0$ b) $x^2 - 11x + 5 = 0$

   c) $x^2 + 5x - 4 = 0$ d) $3x^2 + 11x + 2 = 0$

   e) $x + 2 - \dfrac{5}{x} = 0$ f) $2x^2 = 7 - 4x$

2 Write down the quadratic equation whose roots have the sum and the product given.

   a) Sum 7; product 12   b) Sum -3; product 2

   c) Sum -2; product -4 d) Sum -5; product -11

3 The equation $2x^2 - 7x + q = 0$ has a root of 3. Find

   a) the value of q and b) the other root of the equation.

4 The equation $3x^2 + 5x + t = 0$ has a root of -2. Find

   a) the value of t and b) the other root of the equation.

5 The equation $3x^2 - 10x + 6 = 0$ has roots α and β. Without solving the given equation, find a quadratic equation with integer coefficients whose roots are $(\alpha + \beta)$ and $\alpha\beta$.

6 For a general quadratic equation $ax^2 + bx + c = 0$, the roots are

$$\frac{-b \pm \sqrt{b^2 - 4ac}}{2a}$$

Confirm that the sum of these two roots is $-\dfrac{b}{a}$ and that the

product of these two roots is $\dfrac{c}{a}$.

1.2 Symmetrical functions of roots

You have seen how you can derive a quadratic equation given clues about its roots. Example 3 on pages 3–4 extends this idea to roots that are functions of α and β.

Functions of α and β include:

α^2 and β^2

α^3 and β^3

$(\alpha + 1)$ and $(\beta + 1)$

$2\alpha + \dfrac{3}{\alpha}$ and $2\beta + \dfrac{3}{\beta}$

Symmetrical functions of α and β include:

$\alpha^2 + \beta^2$, $\alpha^2\beta^2$

$\alpha^3 + \beta^3$, $\alpha^3\beta^3$

$(\alpha + 1) + (\beta + 1)$, $(\alpha + 1)(\beta + 1)$

$\left(2\alpha + \dfrac{3}{\alpha}\right) + \left(2\beta + \dfrac{3}{\beta}\right)$, $\left(2\alpha + \dfrac{3}{\alpha}\right)\left(2\beta + \dfrac{3}{\beta}\right)$

> In each case, the functions of α and β mirror each other.

FP1

> These are called symmetrical functions because they remain unchanged if a and b are interchanged. All of these symmetrical functions can be expressed in terms of $\alpha + \beta$ and $\alpha\beta$.

To find a quadratic equation whose roots are functions of α and β, you can start by finding the sum and product of the new roots.

Algebraic techniques you can use include the following:

✦ If the new roots are α^2 and β^2, then their product $\alpha^2 \times \beta^2$ is equal to $(\alpha\beta)^2$ and their sum $\alpha^2 + \beta^2$ is found from:

$$(\alpha + \beta)^2 = \alpha^2 + 2\alpha\beta + \beta^2$$
$$\therefore \quad \alpha^2 + \beta^2 = (\alpha + \beta)^2 - 2\alpha\beta$$

> **Remember**
> You can add fractions by finding a common denominator:
> $$\frac{a}{b} + \frac{c}{d} = \frac{ad + bc}{bd}$$

✦ If the new roots are $\dfrac{1}{\alpha}$ and $\dfrac{1}{\beta}$, then their product is:

$$\frac{1}{\alpha} \times \frac{1}{\beta} = \frac{1}{\alpha\beta}$$

and their sum is:

$$\frac{1}{\alpha} + \frac{1}{\beta} = \frac{\beta + \alpha}{\alpha\beta} \quad \text{or} \quad \frac{\alpha + \beta}{\alpha\beta}.$$

✦ If the new roots are α^3 and β^3, then their product is:

$$\alpha^3 \times \beta^3 = \alpha^3\beta^3 = (\alpha\beta)^3$$

and their sum, $\alpha^3 + \beta^3$, is found by using the cube of $\alpha + \beta$:

$$(\alpha + \beta)^3 = \alpha^3 + 3\alpha^2\beta + 3\alpha\beta^2 + \beta^3$$
$$\therefore \quad \alpha^3 + \beta^3 = (\alpha + \beta)^3 - 3\alpha^2\beta - 3\alpha\beta^2$$
$$= (\alpha + \beta)^3 - 3\alpha\beta(\alpha + \beta)$$

> You could use the **binomial theorem** to cube $\alpha + \beta$. This is explained in Module C2.

Example 4

The equation $3x^2 + 7x - 5 = 0$ has roots α and β.

Find the quadratic equation whose roots are $\dfrac{1}{\alpha}$ and $\dfrac{1}{\beta}$.

From $3x^2 + 7x - 5 = 0$,

$$\alpha + \beta = -\frac{7}{3} \text{ and } \alpha\beta = -\frac{5}{3}$$

The sum of the new roots is:

$$\frac{1}{\alpha} + \frac{1}{\beta} = \frac{\beta + \alpha}{\alpha\beta}$$

$$= \frac{-\dfrac{7}{3}}{-\dfrac{5}{3}}$$

$$= \frac{7}{5}$$

The product of the new roots is:

$$\frac{1}{\alpha} \times \frac{1}{\beta} = \frac{1}{\alpha\beta}$$

$$= \frac{1}{-\dfrac{5}{3}}$$

$$= -\frac{3}{5}$$

Therefore, the new equation is $x^2 - \dfrac{7}{5}x - \dfrac{3}{5} = 0$ or $5x^2 - 7x - 3 = 0$

FP1

Example 5

The equation $4x^2 + 7x - 5 = 0$ has roots α and β. Find the quadratic equation whose roots are α^2 and β^2.

From $4x^2 + 7x - 5 = 0$,

$$\alpha + \beta = -\frac{7}{4} \text{ and } \alpha\beta = -\frac{5}{4}$$

The sum of the new roots is:

$$\alpha^2 + \beta^2 = (\alpha + \beta)^2 - 2\alpha\beta$$

Substituting the above values in the RHS:

$$\alpha^2 + \beta^2 = \left(-\frac{7}{4}\right)^2 - 2\left(-\frac{5}{4}\right) = \frac{89}{16}$$

RHS is an abbreviation for **right-hand side**.

The product of the new roots is $\alpha^2\beta^2 = (\alpha\beta)^2$. Substituting the value for $\alpha\beta$ gives:

$$\alpha^2\beta^2 = \left(-\frac{5}{4}\right)^2 = \frac{25}{16}$$

Therefore, the new equation is:

$$x^2 - \frac{89}{16}x + \frac{25}{16} = 0 \quad \text{or} \quad 16x^2 - 89x + 25 = 0$$

Example 6

Find the quadratic equation whose roots are α^3 and β^3, where α and β are the roots of $x^2 - 5x - 7 = 0$.

From $x^2 - 5x - 7 = 0$,

$$\alpha + \beta = 5 \text{ and } \alpha\beta = -7$$

The sum of the new roots is $\alpha^3 + \beta^3$.

Using $\alpha^3 + \beta^3 = (\alpha + \beta)^3 - 3\alpha\beta(\alpha + \beta)$
$$= 5^3 - 3 \times -7 \times 5$$
$$= 125 + 105$$
$$= 230$$

The expansion of $(\alpha + \beta)^3$ is given on page 5.

The product of the roots is $\alpha^3\beta^3$. Hence:

$$\alpha^3\beta^3 = -343$$

Therefore, the new equation is:

$$x^2 - 230x - 343 = 0$$

Example 7

Find the quadratic equation whose roots are $\alpha + \dfrac{5}{\beta}$ and $\beta + \dfrac{5}{\alpha}$

where α and β are the roots of the equation $2x^2 + 7x - 3 = 0$.

From $2x^2 + 7x - 3 = 0$,

$$\alpha + \beta = -\frac{7}{2} \text{ and } \alpha\beta = -\frac{3}{2}$$

FP1

The sum of the new roots is:

$$\alpha + \frac{5}{\beta} + \beta + \frac{5}{\alpha} = \alpha + \beta + \frac{5}{\beta} + \frac{5}{\alpha}$$

$$= \alpha + \beta + \frac{5\alpha + 5\beta}{\alpha\beta}$$

$$= \alpha + \beta + \frac{5(\alpha + \beta)}{\alpha\beta}$$

$$= \frac{-7}{2} + \frac{5 \times -\frac{7}{2}}{-\frac{3}{2}}$$

$$= \frac{-7}{2} + \frac{35}{3}$$

$$= \frac{49}{6}$$

> Substitute the numerical values of $\alpha + \beta$ and $\alpha\beta$.

The product of the new roots is:

$$(\alpha + \frac{5}{\beta})(\beta + \frac{5}{\alpha}) = \alpha\beta + 5 + 5 + \frac{25}{\alpha\beta}$$

$$= -\frac{3}{2} + 10 - \frac{50}{3}$$

$$= -\frac{49}{6}$$

> Substitute the numerical value of $\alpha\beta$.

Therefore, the new equation is:

$$x^2 - \frac{49}{6}x - \frac{49}{6} = 0 \quad \text{or} \quad 6x^2 - 49x - 49 = 0$$

FP1

Exercise 1B

1 The roots of the equation $2x^2 + 5x - 8 = 0$ are α and β.

Find the quadratic equation whose roots are:

a) 7α and 7β
b) $\frac{1}{\alpha}$ and $\frac{1}{\beta}$

c) α^2 and β^2
d) α^3 and β^3

e) $5\alpha + \frac{1}{\alpha}$ and $5\beta + \frac{1}{\beta}$
f) $7\alpha + \frac{1}{\beta}$ and $7\beta + \frac{1}{\alpha}$

2 The roots of the equation $3x^2 - 5x - 6 = 0$ are α and β.

Find the quadratic equation whose roots are:

a) $\dfrac{2}{\alpha}$ and $\dfrac{2}{\beta}$

b) 4α and 4β

c) α^2 and β^2

d) α^3 and β^3

e) $2\alpha + \dfrac{3}{\alpha}$ and $2\beta + \dfrac{3}{\beta}$

1.3 Equations with related roots

If α and β are the roots of the equation

$$ax^2 + bx + c = 0$$

you can obtain an equation with roots 2α and 2β by making the substitution $y = 2x$.

> The technique described in section 1.3 will not be assessed in your FP1 examination.

FP1

Replace the x each time it occurs by $\frac{1}{2}y$, giving:

$$a(\tfrac{1}{2}y)^2 + b(\tfrac{1}{2}y) + c = 0.$$

If you try the value $y = 2\alpha$ in this equation, you obtain:

$$a\alpha^2 + b\alpha + c = 0.$$

This is clearly true, since α is a root of the original equation

$$ax^2 + bx + c = 0.$$

Similarly, $y = 2\beta$ will satisfy the equation.

Hence the required equation is:

$$a(\tfrac{1}{2}y)^2 + b(\tfrac{1}{2}y) + c = 0$$

which can also be written in terms of x as:

$$a(\tfrac{1}{2}x)^2 + b(\tfrac{1}{2}x) + c = 0$$

or, multiplying throughout by 4 to clear the denominators, as:

$$ax^2 + 2bx + 4c = 0.$$

Example 8

Find the equation whose roots are 3α and 3β, where α and β are the roots of the equation $2x^2 - 5x + 3 = 0$.

$$2x^2 - 5x + 3 = 0$$

replace x by $\dfrac{y}{3}$ to give:

$$2\left(\dfrac{y}{3}\right)^2 - 5\left(\dfrac{y}{3}\right) + 3 = 0$$

$$2y^2 - 15y + 27 = 0$$

If the equation is to be expressed in terms of x,

$$2x^2 - 15x + 27 = 0$$

Example 9

Find a quadratic equation with roots $2\alpha - 1$ and $2\beta - 1$, where α and β are the roots of the equation $4x^2 + 7x - 5 = 0$.

$$4x^2 + 7x - 5 = 0$$

Use the variable y, where $y = 2x - 1$, so that $x = \tfrac{1}{2}(y + 1)$.

The required equation is, therefore:

$$4(\tfrac{1}{2}(y + 1))^2 + 7(\tfrac{1}{2}(y + 1)) - 5 = 0,$$

which can be simplified to:

$$y^2 + \tfrac{11}{2}y - \tfrac{1}{2} = 0 \quad \text{or} \quad 2y^2 + 11y - 1 = 0$$

FP1

Exercise 1C

1   The roots of the equation $x^2 + 7x + 11 = 0$ are α and β. Find the quadratic equation whose roots are 2α and 2β.

2   The roots of the equation $x^2 - 15x + 7 = 0$ are α and β. Find the quadratic equation whose roots are 3α and 3β.

3   The roots of the equation $2x^2 + 3x + 17 = 0$ are α and β. Find the quadratic equation whose roots are $\alpha + 1$ and $\beta + 1$.

4   The roots of the equation $3x^2 - 7x + 15 = 0$ are α and β. Find the quadratic equation whose roots are $\alpha - 5$ and $\beta - 5$.

5   The equation $2x^2 + 7x + 3 = 0$ has roots α and β. Find the quadratic equation whose roots are:

a) $2\alpha, 2\beta$ b) $\dfrac{\alpha}{3}, \dfrac{\beta}{3}$

c) $\alpha + 2, \beta + 2$ d) $2\alpha + 1, 2\beta + 1$

6   The equation $3x^2 + 9x - 2 = 0$ has roots α and β. Find the equation whose roots are:

a) $4\alpha, 4\beta$ b) $\dfrac{\alpha}{2}, \dfrac{\beta}{2}$

c) $\alpha - 3, \beta - 3$ d) $3\alpha - 2, 3\beta - 2$

7   The equation $x + 2 + \dfrac{3}{x} = 0$ has roots α and β. Find the equation whose roots are 5α and 5β.

Summary

You should know how to ...	Check out
1 Find the sum and product of the roots of a quadratic equation.	**1** Find the sum and product of the roots of each of the following quadratic equations: a) $x^2 - 6x + 8 = 0$ b) $2x^2 + 5x - 1 = 0$
2 Evaluate symmetrical functions of these roots.	**2** Given that α and β are the roots of the quadratic equation $3x^2 - x - 2 = 0$, find the values of: a) $(\alpha + \beta)^2$ b) $\alpha^2 + \beta^2$ c) $\alpha^2\beta^2$
3 Form new equations with roots related to those of the original equation.	**3** The equation $x^2 + 4x + 2 = 0$ has roots α and β. Form quadratic equations with the roots: a) 3α and 3β b) $\dfrac{3}{\alpha}$ and $\dfrac{3}{\beta}$

FP1

Revision exercise 1

1 For each of these equations, write down the sum and product of the roots. Then solve each equation to verify that your values for the sum and product are correct.

a) $x^2 - 5x + 6 = 0$ b) $x^2 + x - 6 = 0$

c) $x^2 - 6x = 0$ d) $x^2 - 9 = 0$

2 The quadratic equation $x^2 + 3x - 2 = 0$ has roots α and β.
Without finding the values of α and β, find equations with roots:

a) $-\alpha$ and $-\beta$ b) 2α and 2β c) $\dfrac{1}{\alpha}$ and $\dfrac{1}{\beta}$

3 a) The roots of the quadratic equation $x^2 + 4x - 3 = 0$ are α and β.
Without solving the equation, find the value of:

i) $\alpha^2 + \beta^2$

ii) $\left(\alpha^2 + \dfrac{2}{\beta}\right)\left(\beta^2 + \dfrac{2}{\alpha}\right)$

b) Determine a quadratic equation with integer coefficients which has roots:

$\left(\alpha^2 + \dfrac{2}{\beta}\right)$ and $\left(\beta^2 + \dfrac{2}{\alpha}\right)$.

(AQA, 2002)

4 a) The quadratic equation $2x^2 - 6x + 1 = 0$ has roots α and β.
Write down the numerical values of:
i) $\alpha\beta$
ii) $\alpha + \beta$

b) Another quadratic equation has roots $\dfrac{1}{\alpha}$ and $\dfrac{1}{\beta}$.
Find the numerical values of:

i) $\dfrac{1}{\alpha} \times \dfrac{1}{\beta}$

ii) $\dfrac{1}{\alpha} + \dfrac{1}{\beta}$

c) Hence, or otherwise, find the quadratic equation with roots $\dfrac{1}{\alpha}$
and $\dfrac{1}{\beta}$, writing your answer in the form $x^2 + px + q = 0$.   *(AQA, 2004)*

FP1

5 The roots of the quadratic equation $x^2 - 3x + 1 = 0$ are α and β.

a) Without solving the equation:
i)   Show that $\alpha^2 + \beta^2 = 7$.
ii)   Find the value of $\alpha^3 + \beta^3$.

b) i)   Show that $\alpha^4 + \beta^4 = (\alpha^2 + \beta^2)^2 - 2(\alpha\beta)^2$.
ii)   Hence find the value of $\alpha^4 + \beta^4$.

c) Determine a quadratic equation with integer coefficients which
has roots $(\alpha^3 - \beta)$ and $(\beta^3 - \alpha)$.   *(AQA, 2003)*

6 The roots of the quadratic equation $x^2 + 3x - 2 = 0$ are α and β.

a) Write down the values of $\alpha + \beta$ and $\alpha\beta$.

b) Without solving the equation, find the value of:

i) $\dfrac{1}{\alpha^2} + \dfrac{1}{\beta^2}$

ii) $\left(\alpha - \dfrac{3}{\beta^2}\right)\left(\beta - \dfrac{3}{\alpha^2}\right)$.

b) Determine a quadratic equation with integer coefficients which
has roots

$$\left(\alpha - \dfrac{3}{\beta^2}\right) \text{ and } \left(\beta - \dfrac{3}{\alpha^2}\right).$$ *(AQA, 2003)*

7 The roots of the quadratic equation

$$x^2 - 5x + 3 = 0$$

are α and β. Form a quadratic equation whose roots are $\alpha + 1$ and
$\beta + 1$, giving your answer in the form $x^2 + px + q = 0$, where p and
q are integers to be determined.   *(AQA, 2001)*

8 The roots of the quadratic equation $3x^2 + 4x - 1 = 0$ are
α and β.

a) Without solving the equation, find the values of:
   i) $\alpha^2 + \beta^2$
   ii) $\alpha^3\beta + \beta^3\alpha$

b) Determine a quadratic equation with integer coefficients
which has roots $\alpha^3\beta$ and $\beta^3\alpha$.   *(AQA, 2001)*

9 The quadratic equation

$$x^2 + px + 2 = 0$$

has roots α and β.

a) Write down the value of $\alpha\beta$.

b) Express in terms of p:
   i) $\alpha + \beta$
   ii) $\alpha^2 + \beta^2$

c) Given that $\alpha^2 + \beta^2 = 5$, find the possible values of p.   *(AQA, 2003)*

10 The roots of the quadratic equation $x^2 + (7 + p)x + p = 0$ are
α and β.

a) Write down the value of $\alpha + \beta$ and the value of $\alpha\beta$, in terms
of p.

b) Find the value of $\alpha^2 + \beta^2$ in terms of p.

c) i) Show that $(\alpha - \beta)^2 = p^2 + 10p + 49$.
   ii) Given that α and β differ by 5, find the possible values
   of p.   *(AQA, 2004)*

2 Series

This chapter will show you how to

✦ Use the sigma notation for the sum of a series
✦ Use formulae for the sums of squares and cubes
✦ Evaluate sums of series involving squares and cubes

Before you start

FP1

You should know how to ...	Check in
1 Use common factors to factorise lengthy expressions.	**1** Use common factors to factorise the following expressions: a) $(x + 1)(x + 2)(x + 3) + x(x + 1)(x + 2)$ b) $(x + 1)(x + 2)(x + 3) - x(x + 1)(x + 2)$ c) $2(2x - 1)(2x + 1)(2x + 3) + 2(2x + 3)$ d) $(x + 1)^3(x - 2)^2 - (x + 1)^2(x - 2)^3$
2 Substitute one expression into another.	**2** Find z in terms of x, given that: a) $z = (y + 1)(2y + 1)$ and $y = 2x + 1$ b) $z = y(y^2 + 2)$ and $y = x^2$ c) $z = y^2 + 2y + 1$ and $y = x - 1$

> **Links to Core modules**
> Knowledge of arithmetic series from module C2 would be helpful but is not assumed. Note that the formulae for the sums of squares and cubes do not need to be memorised for examination purposes. Nor do the proofs of these formulae have to be known.

2.1 Summation of series

The natural numbers are 1, 2, 3, 4, 5, 6, ...

When you add these numbers cumulatively, you get:

$$1 = 1$$
$$1 + 2 = 3$$
$$1 + 2 + 3 = 6$$
$$1 + 2 + 3 + 4 = 10 \text{ and so on.}$$

> 1, 3, 6, 10, 15, ... are called the **triangle numbers**.

Each of these sums is an arithmetic series, since there is a common difference between consecutive terms.

Denote the sum to n terms by S_n. Then:

$$S_n = 1 + 2 + 3 + \ldots + (n - 2) + (n - 1) + n \qquad (1)$$

Reversing the order of the terms:

$$S_n = n + (n - 1) + (n - 2) + \ldots + 3 + 2 + 1 \qquad (2)$$

Adding (1) and (2) gives:

$$2S_n = (n + 1) + (n + 1) + (n + 1) + \ldots + (n + 1) + (n + 1) + (n + 1)$$

$$= n(n + 1)$$

Dividing each side by 2 gives:

$$S_n = \frac{n}{2}(n + 1)$$

So, the sum of the first n natural numbers is $\frac{n}{2}(n + 1)$. You can extend this to an arithmetic series with first term a and last term l.

The sum of an arithmetic series with n terms is:

$$\frac{n}{2}(a + l) \quad \text{or} \quad \frac{n}{2}(2n + [n - 1]d)$$

where a is the first term, l is the last term and d is the common difference.

> Karl Friedrich Gauss (1777–1855) is commonly regarded as one of the great mathematicians. As a schoolboy, he added the numbers from 1 to 100 very quickly using this method.

> This formula also gives the nth term of the sequence of triangle numbers.

FP1

Sigma notation

The symbol used for writing 'the sum of' is the Greek capital letter sigma, which is written as Σ.

For example, $\sum\limits_{r=1}^{n}(2r + 7)$ is the mathematical way of writing the sum of the values which $2r + 7$ takes as r changes from $r = 1$ to $r = n$.

So you have:

$$\sum_{r=1}^{5}(2r + 7) = 9 + 11 + 13 + 15 + 17 = 65$$

The lower limit does not need to be 1. For example:

$$\sum_{r=4}^{7} r = 4 + 5 + 6 + 7$$

The first result above can be written in sigma notation.

> The notation Σ is explained in module C2.

> $\sum\limits_{4}^{7} r$ is an abbreviated way of writing $\sum\limits_{r=4}^{7} r$.

> The sum of the first n natural numbers is given by:
> $$\sum_{1}^{n} r = \frac{1}{2}n(n + 1)$$

FP1

Example 1

Find $\displaystyle\sum_1^8 r$.

From the formula $\displaystyle\sum_1^n r = \frac{1}{2}n(n+1)$, using $n = 8$:

$$\sum_1^8 r = \frac{1}{2} \times 8 \times 9 = 36$$

> You can check that
> $1 + 2 + 3 + 4 + 5 + 6 + 7 + 8$
> is indeed 36.

Example 2

Find $\displaystyle\sum_1^5 (r + 7)$.

> In this simple example, you can just add the numbers directly to get 50.

Notice that $\displaystyle\sum_1^5 (r + 7)$ may be split into $\displaystyle\sum_1^5 r + \sum_1^5 7$.

The first term

$$\sum_1^5 r = \frac{1}{2} \times 5(5 + 1) = 15$$

The second term, $\displaystyle\sum_1^5 7$, is the sum of 7 repeated 5 times, which is 35.

Therefore,

$$\sum_1^5 (r + 7) = 15 + 35 = 50$$

> An alternative method is:
> $$\sum_1^5(r+7) = 8 + 9 + 10 + 11 + 12$$
> $$= \sum_8^{12} r$$
> $$= \sum_1^{12} r - \sum_1^7 r$$
> $$= \left(\frac{1}{2} \times 12 \times 13\right)$$
> $$- \left(\frac{1}{2} \times 7 \times 8\right)$$
> $$= 78 - 28 = 50$$

The last example shows that you can split up summations.

For example, you can split $\displaystyle\sum_1^n (4r + 3)$ into its two parts $\displaystyle\sum_1^n (4r) + \sum_1^n (3)$

which are $\displaystyle 4\sum_1^n r + 3\sum_1^n 1$. Now:

$$\sum_1^n 1 = 1 + 1 + 1 + 1 + \ldots + 1 = n$$

So, $\displaystyle\sum_1^n (4r + 3) = 4 \times \frac{1}{2}n(n+1) + 3n$

$$= 2n^2 + 5n$$

> $\displaystyle\sum_1^n 1 = n$ is a useful result that can often be used when manipulating sums of series.

Example 3

Evaluate $\displaystyle\sum_1^8 (r + 11)$.

$$\sum_1^8 (r + 11) = \sum_1^8 r + \sum_1^8 11$$

$$= \frac{1}{2} \times 8 \times 9 + 11 \times 8$$

$$= 36 + 88 = 124$$

> Alternatively, note that:
> $$\sum_1^8(r+11) = 12 + 13 + \ldots + 19$$
> $$= \sum_{12}^{19} r = \sum_1^{19} r - \sum_1^{11} r$$
> $$= \frac{1}{2} \times 19 \times 20$$
> $$- \frac{1}{2} \times 11 \times 12$$
> $$= 190 - 66$$
> $$= 124, \text{ as expected.}$$

Example 4

Find $\displaystyle\sum_{70}^{100}(2r - 5)$.

Split the given expression $\displaystyle\sum_{70}^{100}(2r - 5)$ into:

$$\sum_{70}^{100}(2r - 5) = 2\sum_{70}^{100}r - \sum_{70}^{100}5$$

which gives:

$$\sum_{70}^{100}(2r - 5) = 2\left\{\sum_{1}^{100}r - \sum_{1}^{69}r\right\} - 31 \times 5$$

$$= 2\left\{\left(\frac{1}{2} \times 100 \times 101\right) - \left(\frac{1}{2} \times 69 \times 70\right)\right\} - 31 \times 5$$

$$= 2(5050 - 2415) - 155$$

$$= 5115$$

Alternatively, you could find $(2r - 5)$ using the formula for an arithmetic series:

$$S_n = \frac{n}{2}\left[2a + (n - 1)d\right]$$

where $a = 135$, $n = 31$ and $d = 2$.
Then the sum becomes:

$$S_{31} = \frac{31}{2}(2 \times 135 + 30 \times 2)$$

$$= \frac{31}{2}(270 + 60)$$

$$= 5115, \text{ as expected.}$$

FP1

Exercise 2A

1 Find the values of:

a) $\displaystyle\sum_{r=1}^{10}r$ b) $\displaystyle\sum_{1}^{25}r$ c) $\displaystyle\sum_{1}^{30}(3r + 1)$

d) $\displaystyle\sum_{1}^{7}(5r - 6)$ e) $\displaystyle\sum_{11}^{25}r$ f) $\displaystyle\sum_{15}^{25}(2r + 4)$

2 Find p and q such that $\displaystyle\sum_{1}^{25}(r + 17) = \sum_{p}^{q}r$.

2.2 Sums of the squares of natural numbers

The square numbers are:

$\qquad$ 1, 4, 9, 16, 25, 36, ...

and the cumulative totals of these numbers are:

$\qquad 1 = 1$
$\qquad 1 + 4 = 5$
$\qquad 1 + 4 + 9 = 14$
$\qquad 1 + 4 + 9 + 16 = 30$
$\qquad 1 + 4 + 9 + 16 + 25 = 55$ and so on.

These sums, 1, 5, 14, 30, 55, ..., do not form an obvious pattern.
However, if you multiply each sum by 6, you get 6, 30, 84, 180, 330 ...

6 times the sum of the first square is $1 \times 2 \times 3$
6 times the sum of the first two squares is $2 \times 3 \times 5$
6 times the sum of the first three squares is $3 \times 4 \times 7$
6 times the sum of the first four squares is $4 \times 5 \times 9$

Note This is a **demonstration** of a general result, **not a proof**.

which leads to:

$$6\sum_1^n r^2 = n(n + 1)(2n + 1)$$

and hence:

$$\sum_1^n r^2 = \frac{n}{6}(n + 1)(2n + 1)$$

∴ The sum of the squares of the first n natural numbers is given by:

$$\sum_1^n r^2 = \frac{1}{6}n(n + 1)(2n + 1)$$

Check this result for $n = 4$:
$$\sum_1^4 r^2 = \frac{1}{6} \times 4 \times 5 \times 9 = 30$$

This formula is in the formulae book issued by AQA and can be used in examinations without proof.

FP1

Example 5

Find $\sum_1^8 r^2$.

....

Use the formula $\sum_1^n r^2 = \frac{1}{6}n(n + 1)(2n + 1)$, with $n = 8$:

$$\sum_1^8 r^2 = \frac{1}{6} \times 8 \times 9 \times 17 = 204$$

Example 6

Use the appropriate formula to find the sum of the squares of the natural numbers from 11 to 20.

....

Use $\sum_{11}^{20} r^2 = \sum_1^{20} r^2 - \sum_1^{10} r^2$, which gives:

$$\left(\frac{1}{6} \times 20 \times 21 \times 41\right) - \left(\frac{1}{6} \times 10 \times 11 \times 21\right)$$

$$= 2870 - 385 = 2485$$

Example 7

Find $\sum_1^{10}(5r^2 - 2r)$.

....

$$\sum_1^{10}(5r^2 - 2r) = 5\sum_1^{10} r^2 - 2\sum_1^{10} r$$

$$= 5\left(\frac{1}{6} \times 10 \times 11 \times 21\right) - 2 \times \left(\frac{1}{2} \times 10 \times 11\right)$$

$$= 1925 - 110 = 1815$$

Example 8

Find an expression in n for $\displaystyle\sum_{r=1}^{n}(4r^2+1)$.

First, split the given term into its parts, and then use the formulae as appropriate.

$$\sum_{r=1}^{n}(4r^2+1)=4\sum_{r=1}^{n}r^2+\sum_{r=1}^{n}1$$

which gives:

$$\sum_{r=1}^{n}(4r^2+1)=4\times\frac{1}{6}n(n+1)(2n+1)+n$$

$$=\frac{2n}{3}(n+1)(2n+1)+n=\frac{1}{3}[2n(n+1)(2n+1)+3n]$$

Remember: $\displaystyle\sum_{r=1}^{n}1=n$

FP1

Example 9

Find $\displaystyle\sum_{r=1}^{8}(r^2+2)$.

$$\sum_{r=1}^{8}(r^2+2)=\sum_{1}^{8}r^2+\sum_{1}^{8}2$$

$$\sum_{r=1}^{n}(r^2+2)=\frac{1}{6}n(n+1)(2n+1)+2n$$

Now $n=8$, therefore:

$$\sum_{r=1}^{8}(r^2+2)=\left(\frac{1}{6}\times8\times9\times17\right)+16=220$$

Hence: $\displaystyle\sum_{r=1}^{8}(r^2+2)=220$

Example 10

Find $\displaystyle\sum_{1}^{n}(3r^2+7r-2)$

$$\sum_{1}^{n}(3r^2+7r-2)=\sum_{1}^{n}3r^2+\sum_{1}^{n}7r-\sum_{1}^{n}2$$

$$=3\sum_{1}^{n}r^2+7\sum_{1}^{n}r-\sum_{1}^{n}2$$

$$=3\times\frac{1}{6}\times n(n+1)(2n+1)+\frac{7}{2}n(n+1)-2n$$

$$=\frac{n}{2}\{(n+1)(2n+1)+7(n+1)-4\}$$

$$=\frac{n}{2}\{(2n^2+3n+1+7n+7-4\}$$

$$=\frac{n}{2}(2n^2+10n+4)$$

Therefore: $\displaystyle\sum_{1}^{n}(3r^2+7r-2)=n(n^2+5n+2)$

Example 11

Find the sum of the squares of the odd numbers from 1 to 49.

Sum of the squares of the odd numbers from 1 to 49
  = Sum of the squares of all the numbers from 1 to 49
  − Sum of the squares of the even numbers from 1 to 49

Sum of the squares of the even numbers is:

$$2^2 + 4^2 + 6^2 + \ldots + 48^2$$

$$= 2^2(1^2 + 2^2 + 3^2 + 4^2 + \ldots + 24^2)$$

$$= 4\sum_1^{24} r^2$$

$4 \times$ sum of squares from 1 to 24.

Sum of the squares of the odd numbers from 1 to 49 is:

$$\sum_1^{49} r^2 - 4\sum_1^{24} r^2$$

Sum of squares from 1 to 49 − 4 × sum of squares from 1 to 24.

$$= \left(\frac{1}{6} \times 49 \times 50 \times 99\right) - 4\left(\frac{1}{6} \times 24 \times 25 \times 49\right)$$

$$= 40\,425 - 19\,600 = 20\,825$$

FP1

Exercise 2B

1 Find the values of:

a) $\displaystyle\sum_1^{15} r^2$

b) $\displaystyle\sum_1^{20} 2r^2$

c) $\displaystyle\sum_1^{25} (3r^2 + 1)$

2 Find, in terms of n,

a) $\displaystyle\sum_1^{n} 3r^2$

b) $\displaystyle\sum_1^{n} (4r^2 + 2r + 1)$

c) $\displaystyle\sum_1^{n} (5r^2 + 3r - 6)$

3 Find the sum of the squares of the even numbers between 1 and 41.

4 Find the sum of the squares of the odd numbers between 40 and 60.

2.3 Sums of the cubes of natural numbers

The cube numbers are 1, 8, 27, 64, … .

And the cumulative totals of these numbers are:

$$1 = 1$$
$$1 + 8 = 9$$
$$1 + 8 + 27 = 36$$
$$1 + 8 + 27 + 64 = 100 \text{ and so on.}$$

Notice that these totals are square numbers: $1^2, 3^2, 6^2, 10^2, …$, which are the squares of the triangle numbers.

On page 15, you learnt that the nth term of the sequence of triangle numbers is given by the expression $\frac{1}{2}n(n + 1)$.

Thus, the sum of the cubes of the first n natural numbers is:

$$\left[\frac{1}{2}n(n + 1)\right]^2 \quad \text{or} \quad \frac{1}{4}n^2(n + 1)^2.$$

> The square of the nth triangle number is the sum of the cubes of the first n natural numbers.
> For example, when $n = 4$:
> $$1^3 + 2^3 + 3^3 + 4^3 = 100 = 10^2$$
> 10 is the 4th triangle number.

FP1

$$\sum_1^n r^3 = \frac{1}{4}n^2(n + 1)^2$$

> You can prove the formulae for $\sum_1^n r^3$ and also the formula for $\sum_{r=1}^n r^2$ given on page 18, using the method of induction or that of differencing, which are explained in Module FP2.

Example 12

Evaluate $\displaystyle\sum_1^7 2r^3$.

....

$$\sum_1^7 2r^3 = 2\sum_1^7 r^3 = 2 \times \tfrac{1}{4} \times 7^2 \times 8^2 = 1568$$

Example 13

Find the value of $\displaystyle\sum_{r=n+1}^{2n} (4r^3 - 3)$.

....

$$\sum_{r=n+1}^{2n} (4r^3 - 3) = \sum_{r=1}^{2n} (4r^3 - 3) - \sum_{r=1}^{n} (4r^3 - 3)$$

which gives:

$$\sum_{r=n+1}^{2n} (4r^3 - 3) = 4\sum_1^{2n} r^3 - 3\sum_1^{2n} 1 - \left(4\sum_1^n r^3 - 3\sum_1^n 1\right)$$

$$= 4 \times \frac{1}{4}(2n)^2(2n + 1)^2 - 3 \times 2n - \left[4 \times \frac{1}{4}n^2(n + 1)^2 - 3n\right]$$

$$= 4n^2(2n + 1)^2 - 6n - n^2(n + 1)^2 + 3n$$

$$= 16n^4 + 16n^3 + 4n^2 - n^4 - 2n^3 - n^2 - 3n$$

Therefore:

$$\sum_{r=n+1}^{2n} (4r^3 - 3) = 15n^4 + 14n^3 + 3n^2 - 3n$$

> Subtract $\displaystyle\sum_{r=1}^n$ from $\displaystyle\sum_{r=1}^{2n}$ to get $\displaystyle\sum_{r=n+1}^{2n}$

> Use the formula for the sum of the cubes of $2n$ and n numbers.

> $4 \times \dfrac{1}{4}(2n)^2 = 4n^2$

Example 14

Find $\displaystyle\sum_{r=1}^{n} (2r^3 + 3r^2 + 1)$.

$$\sum_{r=1}^{n} (2r^3 + 3r^2 + 1) = \sum_{r=1}^{n} 2r^3 + \sum_{r=1}^{n} 3r^2 + \sum_{r=1}^{n} 1$$

$$= 2\sum_{r=1}^{n} r^3 + 3\sum_{r=1}^{n} r^2 + \sum_{r=1}^{n} 1$$

which gives:

$$\sum_{r=1}^{n} (2r^3 + 3r^2 + 1) = 2 \times \frac{1}{4}n^2(n+1)^2 + 3 \times \frac{1}{6}n(n+1)(2n+1) + n$$

$$= \frac{n}{2}[n(n+1)^2 + (n+1)(2n+1) + 2]$$

$$= \frac{n}{2}[n(n^2 + 2n + 1) + (2n^2 + 3n + 1) + 2]$$

> Take out the common factor $\frac{n}{2}$.

Therefore:

$$\sum_{r=1}^{n} (2r^3 + 3r^2 + 1) = \frac{n}{2}(n^3 + 4n^2 + 4n + 3)$$

FP1

Exercise 2C

1 Find $\displaystyle\sum_{r=1}^{n} (2r^2 + 2r)$.

2 Find $\displaystyle\sum_{r=1}^{n} (2r^3 + r)$.

3 Find $\displaystyle\sum_{r=1}^{n} (r+1)(r-2)$.

4 Find $\displaystyle\sum_{r=1}^{n} (2r-1)(r+5)$.

5 Given that n is a positive integer, find $\displaystyle\sum_{r=1}^{n} (2r-3)^3$, giving your answer in its simplest form.

6 Show that $\displaystyle\sum_{r=1}^{n} r(2r+1) = \frac{1}{6}n(n+1)(4n+5)$. Hence evaluate

$$\sum_{r=10}^{40} r(2r+1).$$

7 Write down the sum $\displaystyle\sum_{r=1}^{2N} r^3$ in terms of N, and hence find

$$1^3 - 2^3 + 3^3 - 4^3 + \ldots - (2N)^3$$

in terms of N, simplifying your answer.

Summary

You should know how to ...	Check out
1 Use the sigma notation for the sum of a series.	**1** Express the following series using the sigma notation: a) $100 + 101 + 102 + ... + 200$ b) $10^2 + 11^2 + 12^2 + ... + 20^2$ c) $2^3 + 4^3 + 6^3 + ... + 50^3$
2 Use formulae for the sums of squares and cubes.	**2** Find expressions in terms of n for the following sums: a) $\displaystyle\sum_{r=1}^{n} (r^2 + r^3)$ b) $\displaystyle\sum_{r=1}^{n-1} (r^2 + r^3)$ c) $\displaystyle\sum_{r=1}^{2n}$
3 Evaluate the sums of series involving squares and cubes.	**3** Use the formulae for Σr^2 and Σr^3 to evaluate: a) $1^2 + 2^2 + 3^2 + ... + 20^2$ b) $10^2 + 11^2 + 12^2 + ... + 20^2$ c) $2^3 + 4^3 + 6^3 + ... + 50^3$

FP1

Revision exercise 2

1 Write down expressions in terms of n for $\displaystyle\sum_{r=1}^{n} r^2$, $\displaystyle\sum_{r=1}^{n} (r^2 + 1)$, $\displaystyle\sum_{r=1}^{n} (r^2 - 5)$.

2 Write down expressions in terms of n for $\displaystyle\sum_{r=1}^{n} r^3$, $\displaystyle\sum_{r=1}^{n} (r^3 + r)$, $\displaystyle\sum_{r=1}^{n} (r^3 - 5r)$.

3 Write down expressions in terms of n for $\displaystyle\sum_{r=1}^{2n} r^2$ and $\displaystyle\sum_{r=1}^{3n} r^3$.

4 a) Find the value of:

    i) $\displaystyle\sum_{r=1}^{100} r^3$ ii) $\displaystyle\sum_{r=51}^{100} r^3$

  b) Find the sum of the fifty integers from 51 to 100 inclusive.

  c) Hence find the value of $\displaystyle\sum_{r=51}^{100} (r^3 - 6325r)$.   *(AQA, 2003)*

5 a) Find the sum of the integers from 1 to 300 inclusive.

   b) Evaluate:

   $$\sum_{r=1}^{300} (r^2 + r)$$

(AQA, 2001)

6 a) Show that $\displaystyle\sum_{r=1}^{2n} r^2 = \tfrac{1}{3}n(2n+1)(4n+1)$.

   b) Show that $\displaystyle\sum_{r=1}^{n} (2r)^2 = \tfrac{2}{3}n(n+1)(2n+1)$.

   c) Using the results in parts a) and b), or otherwise, show that:

   $$\sum_{r=1}^{n} (2r-1)^2 = \tfrac{1}{3}n(2n+1)(2n-1)$$

FP1

7 a) Write down a formula for $\displaystyle\sum_{r=1}^{n} 3r^2$ in terms of n.

   b) Write down a formula for $\displaystyle\sum_{r=1}^{n} r$ in terms of n.

   c) Using your results in parts a) and b), or otherwise, show that:

   $$\sum_{r=1}^{n} r(3r-1) = n^2(n+1)$$

8 a) Show that:

   $$\sum_{r=1}^{n} 3r(r+1) = n(n+1)(n+2)$$

   b) Deduce the numerical value of $\displaystyle\sum_{r=100}^{999} r(r+1)$.

9 Show that $\displaystyle\sum_{r=1}^{n} (2r^3 - 3r^2 + 2r) = \tfrac{1}{2}n(n^3+1)$.

10 Given that $\displaystyle\sum_{r=1}^{n} (ar^2 + br) = n^2(n+1)$, find the values of a and b.

3 Matrices

This chapter will show you how to

◆ Write matrices using different notations
◆ Add, subtract and multiply matrices
◆ Recognise identity matrices and zero matrices

Before you start

You should know how to ...	Check in
1 Give priority to multiplication over addition and subtraction when working out a numerical calculation (Note: many calculators do this automatically.)	**1** Calculate: a) $1 \times 2 + 3 \times 4 + 5 \times 6$ b) $1 \times 2 \times 3 + 4 \times 5 \times 6$
2 Calculate with directed numbers	**2** Calculate: a) $^-5 - {}^-9$ b) $^-3 \times 13 + {}^-4 \times {}^-7$

> **Links to Core modules**
> This topic does not require any knowledge from modules C1 and C2.

A matrix stores mathematical information in a concise way. The information in a matrix is written down in a rectangular array of rows and columns of terms, called **elements** or **entries**, each of which has its own precise position in the array.

$\begin{pmatrix} 4 \\ 8 \\ 7 \end{pmatrix}$ is a matrix, but its meaning depends on the context.

In football, it could represent the number of goals scored in three different matches. In a shop, it could represent the prices of three different items. In mathematics, it could represent a vector.

> The FP1 module addresses only one important application of matrices, which you will learn in chapter 4.

3.1 Notation

A matrix is often represented by a bold capital letter. For example

$$\mathbf{M} = \begin{pmatrix} 4 & 11 \\ 4 & 4 \end{pmatrix}$$

The order of a matrix

The **order** of a matrix is its shape. For example, the matrix $\begin{pmatrix} 6 & -2 & 7 \\ 4 & 3 & -5 \end{pmatrix}$ has order 2×3, since its elements are arranged in two **rows** and three **columns**.

> When stating the order of a matrix, you must first always give the number of rows, followed by the number of columns.

> When the number of rows and the number of columns are equal, the matrix is called a **square matrix**.

FP1 In module FP1, you need only to consider 2×2 and 2×1 matrices.

A 2×2 matrix is a **square matrix**. For example, $\begin{pmatrix} 4 & 1 \\ 2 & 5 \end{pmatrix}$ is a square matrix.

A 2×1 matrix is a **column matrix**. For example, $\begin{pmatrix} 3 \\ -5 \end{pmatrix}$ is a column matrix.

3.2 Addition and subtraction of matrices

> You can only add or subtract matrices if they have the **same order**.

> You add matrices by adding the elements in corresponding positions, for example:
>
> $$\begin{pmatrix} a & b \\ c & d \end{pmatrix} + \begin{pmatrix} e & f \\ g & h \end{pmatrix} = \begin{pmatrix} a+e & b+f \\ c+g & d+h \end{pmatrix}$$

> When you add a 2×2 matrix to a 2×2 matrix, the answer is a 2×2 matrix.

You add two column matrices in the same way. For example:

$$\begin{pmatrix} p \\ q \end{pmatrix} + \begin{pmatrix} r \\ s \end{pmatrix} = \begin{pmatrix} p+r \\ q+s \end{pmatrix}$$

> A 2×1 matrix plus a 2×1 matrix gives you a 2×1 matrix.

You cannot add a square matrix and a column matrix. They do not have the same order.

You subtract matrices of the same order in a similar way.

Example 1

Add **A** and **B**, where:

$$\mathbf{A} = \begin{pmatrix} 3 & 4 \\ -2 & 8 \end{pmatrix} \text{ and } \mathbf{B} = \begin{pmatrix} 4 & -5 \\ 7 & 6 \end{pmatrix}.$$

....

$$\mathbf{A} + \mathbf{B} = \begin{pmatrix} 3 & 4 \\ -2 & 8 \end{pmatrix} + \begin{pmatrix} 4 & -5 \\ 7 & 6 \end{pmatrix} = \begin{pmatrix} 7 & -1 \\ 5 & 14 \end{pmatrix}$$

Example 2

Subtract $\begin{pmatrix} 4 \\ -5 \end{pmatrix}$ from $\begin{pmatrix} 2 \\ 3 \end{pmatrix}$.

$$\begin{pmatrix} 2 \\ 3 \end{pmatrix} - \begin{pmatrix} 4 \\ -5 \end{pmatrix} = \begin{pmatrix} 2 - 4 \\ 3 - -5 \end{pmatrix}$$

$$= \begin{pmatrix} -2 \\ 8 \end{pmatrix}$$

3.3 Multiplication of matrices

Multiplication by a scalar

A **scalar** is a quantity that has size but but no direction.

FP1

For example, when you multiply the matrix $\begin{pmatrix} a & b \\ c & d \end{pmatrix}$ by k, you get:

$$k\begin{pmatrix} a & b \\ c & d \end{pmatrix} = \begin{pmatrix} ka & kb \\ kc & kd \end{pmatrix}$$

In the same way:

$$k\begin{pmatrix} p \\ q \end{pmatrix} = \begin{pmatrix} kp \\ kq \end{pmatrix}$$

Example 3

Find $3\mathbf{M}$, where $\mathbf{M} = \begin{pmatrix} 2 & -1 \\ 4 & 5 \end{pmatrix}$.

$$3\mathbf{M} = 3\begin{pmatrix} 2 & -1 \\ 4 & 5 \end{pmatrix}$$

$$= \begin{pmatrix} 6 & -3 \\ 12 & 15 \end{pmatrix}$$

Example 4

Find $3\mathbf{A} + 2\mathbf{B}$, where $\mathbf{A} = \begin{pmatrix} 4 & 7 \\ 8 & 1 \end{pmatrix}$ and $\mathbf{B} = \begin{pmatrix} 3 & 2 \\ -1 & -3 \end{pmatrix}$.

$$3\mathbf{A} + 2\mathbf{B} = 3\begin{pmatrix} 4 & 7 \\ 8 & 1 \end{pmatrix} + 2\begin{pmatrix} 3 & 2 \\ -1 & -3 \end{pmatrix}$$

$$= \begin{pmatrix} 12 & 21 \\ 24 & 3 \end{pmatrix} + \begin{pmatrix} 6 & 4 \\ -2 & -6 \end{pmatrix}$$

$$= \begin{pmatrix} 18 & 25 \\ 22 & -3 \end{pmatrix}$$

Example 5

Find $2\mathbf{A} - 5\mathbf{B}$, where $\mathbf{A} = \begin{pmatrix} 8 \\ 2 \end{pmatrix}$ and $\mathbf{B} = \begin{pmatrix} 1 \\ -4 \end{pmatrix}$.

$$2\mathbf{A} - 5\mathbf{B} = 2\begin{pmatrix} 8 \\ 2 \end{pmatrix} - 5\begin{pmatrix} 1 \\ -4 \end{pmatrix}$$

$$= \begin{pmatrix} 16 \\ 4 \end{pmatrix} - \begin{pmatrix} 5 \\ -20 \end{pmatrix}$$

$$= \begin{pmatrix} 11 \\ 24 \end{pmatrix}$$

FP1

Exercise 3A

1 $\mathbf{A} = \begin{pmatrix} 3 \\ 4 \end{pmatrix}$ and $\mathbf{B} = \begin{pmatrix} 2 \\ -5 \end{pmatrix}$. Find:

a) $\mathbf{A} + \mathbf{B}$ b) $3\mathbf{A}$

c) $2\mathbf{B}$ d) $7\mathbf{A} + 2\mathbf{B}$

e) $3\mathbf{A} - 4\mathbf{B}$

2 $\mathbf{A} = \begin{pmatrix} 2 & 4 \\ -1 & 3 \end{pmatrix}$ and $\mathbf{B} = \begin{pmatrix} 4 & -7 \\ 3 & 1 \end{pmatrix}$. Find:

a) $2\mathbf{A}$ b) $3\mathbf{B}$

c) $2\mathbf{A} + 3\mathbf{B}$ d) $3\mathbf{A} - 5\mathbf{B}$

3 $\mathbf{P} = \begin{pmatrix} 2 & 8 \\ 4 & -2 \end{pmatrix}$ and $\mathbf{Q} = \begin{pmatrix} 4 & -1 \\ 3 & 5 \end{pmatrix}$. Find:

a) $5\mathbf{P}$ b) $2\mathbf{P} + \mathbf{Q}$

c) $3\mathbf{P} - \mathbf{Q}$ d) $4\mathbf{P} - 5\mathbf{Q}$

Multiplication of a square matrix by a column matrix

The elements in the matrix $\begin{pmatrix} 4 & 5 \\ 2 & 8 \end{pmatrix}$ can represent any items of information that can be expressed in numerical form.

For example, two restaurants *Le Bistro* and *The Grill* offer their customers a choice of still or sparkling bottled water.

John delivers boxes of bottled water to both restaurants and keeps a record of their orders.

Note Applications such as this are not required for the FP1 module. It is only included here to illustrate matrix multiplication.

	Still water	Sparkling water
Le Bistro	5	8
The Grill	4	7

As the delivery is always in the same order, John writes only the numbers of bottles of water $\begin{pmatrix} 5 & 8 \\ 4 & 7 \end{pmatrix}$ as a matrix.

Still water costs £30 per box and sparkling water costs £40 per box.

John writes the costs as a column matrix $\begin{pmatrix} 30 \\ 40 \end{pmatrix}$.

To find out the cost of the water to *Le Bistro*, John calculates:

    5 bottles of still water at £30 per box   = £150
    8 bottles of sparkling water at £40 per box = £320
    The total cost is £470.

FP1

In matrix form:

$$\begin{pmatrix} 5 & 8 \\ \blacksquare & \blacksquare \end{pmatrix}\begin{pmatrix} 30 \\ 40 \end{pmatrix} = \begin{pmatrix} 5 \times 30 + 8 \times 40 \\ \rule{2cm}{0.3em} \end{pmatrix}$$

$$= \begin{pmatrix} 150 + 320 \\ \rule{1.5cm}{0.3em} \end{pmatrix}$$

$$= \begin{pmatrix} 470 \\ \rule{0.8cm}{0.3em} \end{pmatrix}$$

Similarly, for *The Grill*:

$$\begin{pmatrix} \blacksquare & \blacksquare \\ 4 & 7 \end{pmatrix}\begin{pmatrix} 30 \\ 40 \end{pmatrix} = \begin{pmatrix} \rule{2cm}{0.3em} \\ 4 \times 30 + 7 \times 40 \end{pmatrix}$$

$$= \begin{pmatrix} \rule{1.5cm}{0.3em} \\ 120 + 280 \end{pmatrix}$$

$$= \begin{pmatrix} \rule{0.8cm}{0.3em} \\ 400 \end{pmatrix}$$

These two calculations can be combined as

$$\begin{pmatrix} 5 & 8 \\ 4 & 7 \end{pmatrix}\begin{pmatrix} 30 \\ 40 \end{pmatrix} = \begin{pmatrix} 5 \times 30 + 8 \times 40 \\ 4 \times 30 + 7 \times 40 \end{pmatrix}$$

$$= \begin{pmatrix} 470 \\ 400 \end{pmatrix}$$

When you multiply a square matrix $\begin{pmatrix} a & b \\ c & d \end{pmatrix}$ by a column matrix $\begin{pmatrix} p \\ q \end{pmatrix}$, you get a column matrix:

$$\begin{pmatrix} a & b \\ c & d \end{pmatrix}\begin{pmatrix} p \\ q \end{pmatrix} = \begin{pmatrix} ap + bq \\ cp + dq \end{pmatrix}$$

Multiply each row of the square matrix by the column in the second matrix:

$$\begin{pmatrix} a & b \\ c & d \end{pmatrix}\begin{pmatrix} p \\ q \end{pmatrix} = \begin{pmatrix} ap + bq \\ cp + dq \end{pmatrix}$$

Example 6

Multiply $\begin{pmatrix} 3 & 4 \\ 8 & -2 \end{pmatrix}$ by $\begin{pmatrix} 5 \\ -7 \end{pmatrix}$.

$$\begin{pmatrix} 3 & 4 \\ 8 & -2 \end{pmatrix}\begin{pmatrix} 5 \\ -7 \end{pmatrix} = \begin{pmatrix} 3 \times 5 + & 4 \times -7 \\ 8 \times 5 + & -2 \times -7 \end{pmatrix}$$

$$= \begin{pmatrix} 15 - 28 \\ 40 + 14 \end{pmatrix}$$

$$= \begin{pmatrix} -13 \\ 54 \end{pmatrix}$$

FP1

Exercise 3B

1 In each case, multiply the square matrix by the column matrix.

a) $\begin{pmatrix} 1 & 4 \\ 7 & 5 \end{pmatrix}, \begin{pmatrix} 5 \\ 2 \end{pmatrix}$ b) $\begin{pmatrix} 3 & -1 \\ 4 & 2 \end{pmatrix}, \begin{pmatrix} 2 \\ 3 \end{pmatrix}$ c) $\begin{pmatrix} 2 & 0 \\ 5 & -7 \end{pmatrix}, \begin{pmatrix} 2 \\ -1 \end{pmatrix}$.

2 Calculate these matrix products.

a) $\begin{pmatrix} 4 & 8 \\ -1 & 2 \end{pmatrix}\begin{pmatrix} 5 \\ -2 \end{pmatrix}$ b) $\begin{pmatrix} 3 & -7 \\ 4 & 2 \end{pmatrix}\begin{pmatrix} -3 \\ 2 \end{pmatrix}$ c) $\begin{pmatrix} 4 & 5 \\ 0 & 3 \end{pmatrix}\begin{pmatrix} 2 \\ -3 \end{pmatrix}$.

Multiplication of two square matrices

In the scenario on pages 28–29, the cost of the water to the two restaurants was equal to the product

$$\begin{pmatrix} 5 & 8 \\ 4 & 7 \end{pmatrix}\begin{pmatrix} 30 \\ 40 \end{pmatrix}.$$

If the cost of the water is altered to £32 per box for still water and £37 per box for sparkling water, you can write the old and new costs in a square matrix:

$$\begin{pmatrix} 30 & 32 \\ 40 & 37 \end{pmatrix}$$

You work out the cost for the two restaurants at the old prices by multiplying the matrix of the bottle numbers by the price matrix:

$$\begin{pmatrix} 5 & 8 \\ 4 & 7 \end{pmatrix}\begin{pmatrix} 30 & \blacksquare \\ 40 & \blacksquare \end{pmatrix} = \begin{pmatrix} 5 \times 30 + 8 \times 40 & \blacksquare \\ 4 \times 30 + 7 \times 40 & \blacksquare \end{pmatrix}$$

$$= \begin{pmatrix} 470 & \blacksquare \\ 400 & \blacksquare \end{pmatrix}$$

You work out the new cost by multiplying the matrix of the bottle numbers by the **new** price matrix:

$$\begin{pmatrix} 5 & 8 \\ 4 & 7 \end{pmatrix}\begin{pmatrix} \blacksquare & 32 \\ \blacksquare & 37 \end{pmatrix} = \begin{pmatrix} \blacksquare & 5 \times 32 + 8 \times 37 \\ \blacksquare & 4 \times 32 + 7 \times 37 \end{pmatrix}$$

$$= \begin{pmatrix} \blacksquare & 160 + 296 \\ \blacksquare & 128 + 259 \end{pmatrix}$$

$$= \begin{pmatrix} \blacksquare & 456 \\ \blacksquare & 387 \end{pmatrix}$$

You can combine the two calculations:

$$\begin{pmatrix} 5 & 8 \\ 4 & 7 \end{pmatrix}\begin{pmatrix} 30 & 32 \\ 40 & 37 \end{pmatrix} = \begin{pmatrix} 470 & 456 \\ 400 & 387 \end{pmatrix}$$ ← Old and new cost for *Le Bistro*
← Old and new cost for *The Grill*

Number Old New
of bottles price price

FP1

When you multiply two square matrices, $\begin{pmatrix} a & b \\ c & d \end{pmatrix}$ and $\begin{pmatrix} p & q \\ r & s \end{pmatrix}$, you get another square matrix:

Row 1→ $\begin{pmatrix} a & b \\ c & d \end{pmatrix}\begin{pmatrix} p & q \\ r & s \end{pmatrix} = \begin{pmatrix} ap + br & aq + bs \\ cp + dr & cq + ds \end{pmatrix}$

Column 1

Multiply each row in the first matrix by each column in the second matrix:

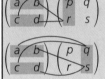

Example 7

Find **AB** and **BA**, where $\mathbf{A} = \begin{pmatrix} 3 & 4 \\ 2 & 5 \end{pmatrix}$ and $\mathbf{B} = \begin{pmatrix} 2 & -1 \\ 6 & 7 \end{pmatrix}$.

$$\mathbf{AB} = \begin{pmatrix} 3 & 4 \\ 2 & 5 \end{pmatrix}\begin{pmatrix} 2 & -1 \\ 6 & 7 \end{pmatrix}$$

$$= \begin{pmatrix} 3 \times 2 + 4 \times 6 & 3 \times -1 + 4 \times 7 \\ 2 \times 2 + 5 \times 6 & 2 \times -1 + 5 \times 7 \end{pmatrix}$$

$$= \begin{pmatrix} 30 & 25 \\ 34 & 33 \end{pmatrix}$$

$$\mathbf{BA} = \begin{pmatrix} 2 & -1 \\ 6 & 7 \end{pmatrix}\begin{pmatrix} 3 & 4 \\ 2 & 5 \end{pmatrix}$$

$$= \begin{pmatrix} 2 \times 3 + -1 \times 2 & 2 \times 4 + -1 \times 5 \\ 6 \times 3 + 7 \times 2 & 6 \times 4 + 7 \times 5 \end{pmatrix}$$

$$= \begin{pmatrix} 4 & 3 \\ 32 & 59 \end{pmatrix}$$

If you change the order in which you multiply the matrices, you generally get a different answer, but *not always*.

3.4 The identity matrix

A square matrix has two diagonals:

$$\begin{pmatrix} a & b \\ c & d \end{pmatrix}$$

a and d are on the **leading diagonal**.

If all the elements on the leading diagonal of a square matrix are equal to 1 and all the other elements are zero, the matrix is an **identity** matrix.

> The 2×2 identity matrix is called **I**, where:
>
> $$\mathbf{I} = \begin{pmatrix} 1 & 0 \\ 0 & 1 \end{pmatrix}$$

FP1

The next example shows you why **I** is called the identity matrix.

Example 8

The matrix $\mathbf{A} = \begin{pmatrix} 2 & -3 \\ 1 & 4 \end{pmatrix}$ and **I** is the identity matrix $\begin{pmatrix} 1 & 0 \\ 0 & 1 \end{pmatrix}$.

Find **AI** and **IA**.

$$\mathbf{AI} = \begin{pmatrix} 2 & -3 \\ 1 & 4 \end{pmatrix}\begin{pmatrix} 1 & 0 \\ 0 & 1 \end{pmatrix}$$

$$= \begin{pmatrix} 2 \times 1 + -3 \times 0 & 2 \times 0 + -3 \times 1 \\ 1 \times 1 + 4 \times 0 & 1 \times 0 + 4 \times 1 \end{pmatrix}$$

$$= \begin{pmatrix} 2 & -3 \\ 1 & 4 \end{pmatrix}$$

$$= \mathbf{A}$$

and $\mathbf{IA} = \begin{pmatrix} 1 & 0 \\ 0 & 1 \end{pmatrix}\begin{pmatrix} 2 & -3 \\ 1 & 4 \end{pmatrix}$

$$= \begin{pmatrix} 1 \times 2 + 0 \times 1 & 1 \times -3 + 0 \times 4 \\ 0 \times 2 + 1 \times 1 & 0 \times -3 + 1 \times 4 \end{pmatrix}$$

$$= \begin{pmatrix} 2 & -3 \\ 1 & 4 \end{pmatrix}$$

$$= \mathbf{A}$$

So $\mathbf{AI} = \mathbf{IA} = \mathbf{A}$

> When a matrix is multiplied by the identity matrix **I**, it stays the same.

The zero matrix

The 2×2 zero matrix, denoted by **O**, is defined by

$$\mathbf{O} = \begin{pmatrix} 0 & 0 \\ 0 & 0 \end{pmatrix}$$

When you multiply **O** by any square matrix **M**, the matrix **O** behaves like the number 0. That is:

$$\mathbf{OM} = \mathbf{MO} = \mathbf{O}$$

> Note that the zero matrix will not be examined.

Exercise 3C

1 $\mathbf{A} = \begin{pmatrix} 4 & 2 \\ 3 & 8 \end{pmatrix}$ and $\mathbf{B} = \begin{pmatrix} 2 & 1 \\ 4 & 5 \end{pmatrix}$.

Find: a) **AB**   b) **BA**

FP1

2 $\mathbf{P} = \begin{pmatrix} -1 & 4 \\ 2 & -3 \end{pmatrix}$ and $\mathbf{Q} = \begin{pmatrix} 5 & 2 \\ 1 & 4 \end{pmatrix}$.

a) Find: i) **PQ**   ii) **QP**   iii) $\mathbf{P}^2$ (that is, **PP**)

b) What do you notice about your answers to a) i) and ii)?

3 $\mathbf{A} = \begin{pmatrix} 3 & 2 \\ -1 & 4 \end{pmatrix}$ and $\mathbf{B} = \begin{pmatrix} 2 & 3 \\ 5 & 0 \end{pmatrix}$.

Find: a) **AB**   b) **BA**

4 $\mathbf{C} = \begin{pmatrix} 2 & 0 \\ 0 & 4 \end{pmatrix}$ and $\mathbf{D} = \begin{pmatrix} 0 & -5 \\ 3 & 0 \end{pmatrix}$.

Find: a) **CD**   b) **DC**

5 $\mathbf{M} = \begin{pmatrix} 2 & 0 \\ 0 & 2 \end{pmatrix}$.

Find: a) $\mathbf{M}^2$ b) $\mathbf{M}^3$ c) $\mathbf{M}^4$

6 $\mathbf{A} = \begin{pmatrix} 3 & 0 \\ 0 & 3 \end{pmatrix}$, $\mathbf{B} = \begin{pmatrix} \dfrac{1}{2} & \dfrac{\sqrt{3}}{2} \\ -\dfrac{\sqrt{3}}{2} & \dfrac{1}{2} \end{pmatrix}$ and $\mathbf{C} = \begin{pmatrix} 0 & 1 \\ -1 & 0 \end{pmatrix}$.

Find: a) **AC**   b) **CA**   c) $\mathbf{B}^2$
      d) $\mathbf{B}^3$ e) $\mathbf{B}^6$ f) $\mathbf{AB}^2$
      g) **AB**   h) **BC**

7 Given that $\mathbf{M} = \begin{pmatrix} 2 & 5 \\ -3 & 4 \end{pmatrix}$, write down the matrices:

a) **IM**   b) **MI**   c) **OM**   d) **MO**

Summary

You should know how to ...	Check out
1 Add, subtract and multiply matrices.	**1** Calculate: a) $\begin{pmatrix} 1 & 3 \\ 4 & 5 \end{pmatrix} + \begin{pmatrix} 3 & 1 \\ 5 & 4 \end{pmatrix} - \begin{pmatrix} 3 & 4 \\ 9 & 8 \end{pmatrix}$ b) $\begin{pmatrix} 1 & 2 \\ 3 & 4 \end{pmatrix}\begin{pmatrix} 5 \\ 6 \end{pmatrix}$ c) $\begin{pmatrix} 1 & 0 \\ 3 & 1 \end{pmatrix}\begin{pmatrix} 1 & 3 \\ 0 & 1 \end{pmatrix}$
2 Recognise identity matrices and zero matrices.	**2** Given that $\mathbf{A} = \begin{pmatrix} 1 & 2 \\ 3 & 4 \end{pmatrix}$, write down these matrix products a) $\mathbf{AI}$ b) $\mathbf{IA}$ c) $\mathbf{AO}$ d) $\mathbf{OA}$

FP1

Revision exercise 3

1 The matrix $\mathbf{M}$ is given by $\mathbf{M} = \begin{pmatrix} 1 & 3 \\ 0 & 2 \end{pmatrix}$, and $\mathbf{I}$ is the identity matrix $\begin{pmatrix} 1 & 0 \\ 0 & 1 \end{pmatrix}$.

Calculate these matrices: a) $\mathbf{M} - \mathbf{I}$ b) $\mathbf{M}(\mathbf{M} - \mathbf{I})$ c) $\mathbf{M}^2 - \mathbf{M}$

2 The matrices $\mathbf{A}$ and $\mathbf{B}$ are given by $\mathbf{A} = \begin{pmatrix} 2 & 0 \\ 0 & 2 \end{pmatrix}$, $\mathbf{B} = \begin{pmatrix} 3 & -1 \\ 1 & 5 \end{pmatrix}$.

Calculate these matrices: a) $\mathbf{B}\begin{pmatrix} 1 \\ 3 \end{pmatrix}$ b) $\mathbf{B}\begin{pmatrix} 2 \\ 6 \end{pmatrix}$ c) $\mathbf{AB}\begin{pmatrix} 1 \\ 3 \end{pmatrix}$

3 The matrices $\mathbf{A}$, $\mathbf{B}$ and $\mathbf{C}$ are given by $\mathbf{A} = \begin{pmatrix} 4 & 3 \\ -2 & 6 \end{pmatrix}$, $\mathbf{B} = \begin{pmatrix} 2 & 4 \\ 1 & 3 \end{pmatrix}$,

$\mathbf{C} = \begin{pmatrix} -3 & 0 \\ 2 & 2 \end{pmatrix}$.

Calculate these matrices: a) $\mathbf{AB}$ b) $\mathbf{AC}$ c) $\mathbf{A}(\mathbf{B} + \mathbf{C})$

4 The matrices $\mathbf{A}$, $\mathbf{B}$ and $\mathbf{C}$ are given by $\mathbf{A} = \begin{pmatrix} 2 & 1 \\ -1 & 4 \end{pmatrix}$, $\mathbf{B} = \begin{pmatrix} 1 & 2 \\ 0 & 3 \end{pmatrix}$,

$\mathbf{C} = \begin{pmatrix} 0 & 2 \\ 1 & -2 \end{pmatrix}$.

Calculate these matrices: a) $\mathbf{AB}$ b) $\mathbf{BC}$ c) $\mathbf{A}(\mathbf{BC})$

                     d) $(\mathbf{AB})\mathbf{C}$

5 The matrices **A** and **B** are given by $\mathbf{A} = \begin{pmatrix} 2 & 0 \\ 0 & 3 \end{pmatrix}$, $\mathbf{B} = \begin{pmatrix} p & q \\ r & s \end{pmatrix}$.

Calculate these matrices: a) **AB**   b) **BA**

6 The matrices **A** and **B** are given by $\mathbf{A} = \begin{pmatrix} 2 & -3 \\ 3 & 2 \end{pmatrix}$, $\mathbf{B} = \begin{pmatrix} -1 & 2 \\ -2 & -1 \end{pmatrix}$.

Calculate these matrices: a) **AB**   b) **BA**

7 The matrices **A** and **B** are given by $\mathbf{A} = \begin{pmatrix} 1 & 2 \\ 3 & 2 \end{pmatrix}$, $\mathbf{B} = \begin{pmatrix} 1 & 1 \\ 2 & 1 \end{pmatrix}$.

a) Calculate the matrices **AB** and **BA**.

b) Show that $(\mathbf{A} + \mathbf{B})^2$ is not equal to $\mathbf{A}^2 + 2\mathbf{AB} + \mathbf{B}^2$.

c) Show that $(\mathbf{A} + \mathbf{B})(\mathbf{A} - \mathbf{B})$ is not equal to $\mathbf{A}^2 - \mathbf{B}^2$.

FP1

8 The matrices **A** and **B** are given by $\mathbf{A} = \begin{pmatrix} p & -q \\ q & p \end{pmatrix}$, $\mathbf{B} = \begin{pmatrix} r & -s \\ s & r \end{pmatrix}$.

Show that $\mathbf{AB} = \mathbf{BA} = \begin{pmatrix} x & -y \\ y & x \end{pmatrix}$, where x and y are to be found in

terms of p, q, r and s.

9 The matrices **A** and **B** are given by $\mathbf{A} = \begin{pmatrix} p & q \\ q & -p \end{pmatrix}$, $\mathbf{B} = \begin{pmatrix} r & s \\ s & -r \end{pmatrix}$.

Show that $\mathbf{AB} = \mathbf{BA} = \begin{pmatrix} x & -y \\ y & x \end{pmatrix}$, where x and y are to be found in

terms of p, q, r and s.

10 The matrices **A** and **B** are given by $\mathbf{A} = \begin{pmatrix} p & -q \\ q & p \end{pmatrix}$, $\mathbf{B} = \begin{pmatrix} r & s \\ s & -r \end{pmatrix}$.

a) Show that $\mathbf{AB} = \begin{pmatrix} x & y \\ y & -x \end{pmatrix}$, where x and y are to be found in

terms of p, q, r and s.

b) Investigate whether it is possible for **AB** to be equal to **BA** in this case.

4 Transformations

This chapter will show you how to

✦ Write certain sines and cosines in surd form
✦ Represent a transformation by a matrix
✦ Identify a transformation from its matrix
✦ Combine two transformations using their matrices

Before you start

You should know how to ...	Check in
1 Identify rotations, reflections, stretches and enlargements.	**1** Find the coordinates of the image of the point $(1, 2)$ under each of the following transformations: a) A rotation through $90°$ anticlockwise about the origin. b) A reflection in the x-axis. c) A reflection in the line $y = x$. d) A stretch parallel to the y-axis with scale factor 2. e) An enlargement with centre the origin and scale factor 2.
2 Use sines and cosines in right-angled triangles.	**2** a) A right-angled triangle has an hypotenuse of 1 and one of the angles is $20°$. Find the lengths of the other two sides to three significant figures. b) A right-angled triangle has an hypotenuse of 10 and one of the angles is $65°$. Find the lengths of the other two sides to three significant figures.
3 Multiply matrices.	**3** Calculate: a) $\begin{pmatrix} 1 & 0 \\ 0 & -1 \end{pmatrix} \begin{pmatrix} 1 \\ 2 \end{pmatrix}$ b) $\begin{pmatrix} 3 & 0 \\ 0 & 1 \end{pmatrix} \begin{pmatrix} 0 & -1 \\ 1 & 0 \end{pmatrix}$.

> **Links to Core modules**
> This chapter refers to sin θ and cos θ, which are introduced at GCSE level and expanded in module C2. It is essential to have covered the material in Chapter 3 before studying this chapter.

You can use a square matrix to define a geometrical transformation in a plane.

The transformation **T** represented by the matrix **M** is defined by:

$$T(x, y) = M\begin{pmatrix} x \\ y \end{pmatrix}.$$

That is, the image of the point (x, y) under the transformation **T** is (x_1, y_1), where $\begin{pmatrix} x_1 \\ y_1 \end{pmatrix} = M\begin{pmatrix} x \\ y \end{pmatrix}.$

For example, if **M** is $\begin{pmatrix} 1 & -1 \\ 1 & 1 \end{pmatrix}$, then the image of the point $(2, 3)$ is found by calculating the matrix product:

$$\begin{pmatrix} 1 & -1 \\ 1 & 1 \end{pmatrix}\begin{pmatrix} 2 \\ 3 \end{pmatrix} = \begin{pmatrix} -1 \\ 5 \end{pmatrix}, \text{ so the image is } (-1, 5).$$

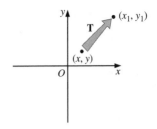

> Note that the image of (0, 0) is always (0, 0).
> The origin is always an **invariant point** in a transformation defined by a matrix.

FP1

4.1 Linear transformations

A linear transformation maps any shape in the x–y plane onto another shape in the plane.

Common linear transformations are:

✦ rotations about the origin
✦ reflections in a line through the origin
✦ stretches parallel to the x- and y-axes
✦ enlargements with centre at the origin

All of these transformations leave the origin invariant.

It is helpful to sketch a diagram, and you should use the same scale on each axis to see the effect of the transformation.

> Every linear transformation can be represented by a 2×2 square matrix, **M**, of the form $\begin{pmatrix} a & b \\ c & d \end{pmatrix}.$

To find the image of the point $(1, 0)$ as a result of the transformation **T**, represented by the matrix **M**, you calculate:

$$M\begin{pmatrix} 1 \\ 0 \end{pmatrix} = \begin{pmatrix} a & b \\ c & d \end{pmatrix}\begin{pmatrix} 1 \\ 0 \end{pmatrix} = \begin{pmatrix} a \\ c \end{pmatrix}$$

So, under **T**, the image of the point $(1, 0)$ is (a, c), which you can see is the first column of the matrix **M**.

Similarly,

$$M\begin{pmatrix} 0 \\ 1 \end{pmatrix} = \begin{pmatrix} a & b \\ c & d \end{pmatrix}\begin{pmatrix} 0 \\ 1 \end{pmatrix} = \begin{pmatrix} b \\ d \end{pmatrix}$$

The calculation tells you that the image of the point $(0, 1)$ is (b, d), which is the second column of **M**.

FP1

4.2 Finding the matrix representing a transformation

To find the matrix **M**, representing a given transformation, you find the images of the two points $(1, 0)$ and $(0, 1)$.

Remember that for the matrix **M** the image of $(1, 0)$ is the first column and the image of $(0, 1)$ is the second column of **M**.

Example 1

Find the matrix, **M**, representing an enlargement, scale factor 2, with the origin as the centre of enlargement.

The images of $(1, 0)$ and $(0, 1)$ are $(2, 0)$ and $(0, 2)$.

So, $\mathbf{M} = \begin{pmatrix} 2 & 0 \\ 0 & 2 \end{pmatrix}$.

You can use the image of $(1, 1)$ to **check**:

$\begin{pmatrix} 2 & 0 \\ 0 & 2 \end{pmatrix}\begin{pmatrix} 1 \\ 1 \end{pmatrix} = \begin{pmatrix} 2 \\ 2 \end{pmatrix}$, as expected.

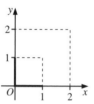

The diagram shows the **unit square** and its image after the enlargement.

Example 2

Find the matrix, **M**, representing a reflection in the line $y = x$.

The images of $(1, 0)$ and $(0, 1)$ are $(0, 1)$ and $(1, 0)$.

So, $\mathbf{M} = \begin{pmatrix} 0 & 1 \\ 1 & 0 \end{pmatrix}$.

Check that the image of $(1, 1)$ is $(1, 1)$.

Example 3

Find the matrix, **M**, representing a stretch parallel to the y-axis with scale factor 4.

The images of $(1, 0)$ and $(0, 1)$ are $(1, 0)$ and $(0, 4)$.

So, $\mathbf{M} = \begin{pmatrix} 1 & 0 \\ 0 & 4 \end{pmatrix}$.

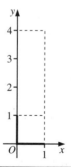

Check that the image of $(1, 1)$ is $(1, 4)$.

Example 4

Find the matrix, **M**, of an anticlockwise rotation about the origin through 25°.

· ·

As the diagram shows, the image of the point $(1, 0)$ is $(\cos 25°, \sin 25°)$ and the image of the point $(0, 1)$ is $(-\sin 25°, \cos 25°)$. These pairs of coordinates become the columns of the matrix, so:

$$\mathbf{M} = \begin{pmatrix} \cos 25° & -\sin 25° \\ \sin 25° & \cos 25° \end{pmatrix}$$

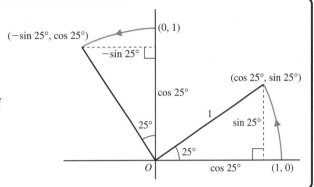

The result of Example 4 can be generalised to give:

The general form for the matrix of a rotation about the origin through angle θ anticlockwise is:

$$\begin{pmatrix} \cos \theta & -\sin \theta \\ \sin \theta & \cos \theta \end{pmatrix}$$

This formula is given in the AQA formulae book.

FP1

Example 5

Find the matrix, **M**, of a reflection in the line $y = (\tan 25°)x$.

· ·

This line has a gradient $\tan 25°$ and so makes an angle 25° with the positive x-axis.

As the diagram shows, the image of the point $(1, 0)$ is $(\cos 50°, \sin 50°)$ and the image of the point $(0, 1)$ is $(\sin 50°, -\cos 50°)$. These pairs of coordinates become the columns of the matrix:

$$\mathbf{M} = \begin{pmatrix} \cos 50° & \sin 50° \\ \sin 50° & -\cos 50° \end{pmatrix}$$

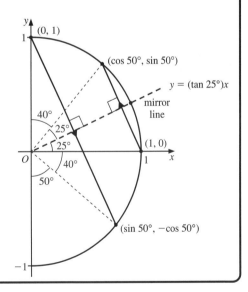

The result of Example 5 can be generalised to give:

The general form for the matrix of a reflection in the line $y = (\tan \theta)x$ is

$$\begin{pmatrix} \cos 2\theta & \sin 2\theta \\ \sin 2\theta & -\cos 2\theta \end{pmatrix}.$$

This formula is given in the AQA formulae book.

Exercise 4A

In Questions **1** to **11**, find the matrix representing the given transformation.

1 An enlargement, scale factor 4, with the origin as the centre of enlargement.

2 A reflection in the *x*-axis.

3 A reflection in the *y*-axis.

4 A reflection in the line $y = -x$.

5 A stretch parallel to the *y*-axis with scale factor 2.

6 A stretch parallel to the *x*-axis with scale factor 3.

7 A rotation about the origin through 90° anticlockwise.

8 A rotation about the origin through 20° anticlockwise.

9 A rotation about the origin through 40° clockwise.

10 A reflection in the line $y = (\tan 20°)\, x$.

11 A reflection in the line $y = (-\tan 20°)\, x$.

> You can check your answers by drawing a diagram and finding the images of (1, 0) and (0, 1) for each transformation.

4.3 Exact values of trigonometric ratios

Examples 4 and 5 show that you use the sines, cosines and tangents of angles when finding the matrices of rotations and reflections. When the angle is 30°, 45° or 60°, you may use exact expressions for the sines, cosines and tangents, rather than decimal approximations.

The angle 45°

ABC is an isosceles right-angled triangle with sides AB and BC each 1 cm.

So, $AC^2 = AB^2 + BC^2$

$\qquad = 1 + 1$

$\qquad = 2$

Therefore, $AC = \sqrt{2}$ cm.

Angle $BAC = 45°$ or $\dfrac{\pi^c}{4}$

Therefore, in triangle ABC

$$\sin A = \frac{1}{\sqrt{2}} \qquad \cos A = \frac{1}{\sqrt{2}} \qquad \tan A = 1$$

Hence:

$$\sin 45° = \frac{1}{\sqrt{2}} \qquad \cos 45° = \frac{1}{\sqrt{2}} \qquad \tan 45° = 1$$

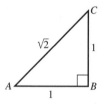

> The angles of a triangle total 180°. $\hat{B} = 90°$ and $\hat{A} = \hat{C}$.

The angles 30° and 60°

ABC is an equilateral triangle with sides 2 cm.

Draw *AD*, the perpendicular bisector of *BC*.

The triangle *ABD* has side *AB* = 2 cm.

As *D* is the mid-point of *BC*, *BD* = 1 cm.

Using Pythagoras' theorem, *AD* = √3 cm.

As *ABC* is an equilateral triangle, angle *B* = 60°. Angle *ADB* = 90° and so angle *BAD* = 30°.

Therefore, in triangle *ABD*

$$\sin 30° = \frac{1}{2} \qquad \cos 30° = \frac{\sqrt{3}}{2} \qquad \tan 30° = \frac{1}{\sqrt{3}}$$

$$\sin 60° = \frac{\sqrt{3}}{2} \qquad \cos 60° = \frac{1}{2} \qquad \tan 60° = \sqrt{3}$$

FP1

You also know that:

$$\sin 0° = 0 \qquad \cos 0° = 1 \qquad \tan 0° = 0$$
$$\sin 90° = 1 \qquad \cos 90° = 0 \qquad \tan 90° \text{ is undefined.}$$

The trigonometric ratios for the angles 0°, 30°, 45°, 60° and 90° are summarised in this table.

Learn the exact values of the trigonometric ratios for these angles.

Angle θ	$\sin \theta$	$\cos \theta$	$\tan \theta$
0°	0	1	0
30°	$\frac{1}{2}$	$\frac{\sqrt{3}}{2}$	$\frac{1}{\sqrt{3}}$
45°	$\frac{1}{\sqrt{2}}$	$\frac{1}{\sqrt{2}}$	1
60°	$\frac{\sqrt{3}}{2}$	$\frac{1}{2}$	$\sqrt{3}$
90°	1	0	Undefined

Example 6

Find the matrix of a rotation about the origin through 45° anticlockwise.

Use the general form for the matrix of a rotation:

$$\mathbf{M} = \begin{pmatrix} \cos\theta & -\sin\theta \\ \sin\theta & \cos\theta \end{pmatrix}$$

Replace the cosines and sines by their exact forms to obtain:

$$\mathbf{M} = \begin{pmatrix} \dfrac{1}{\sqrt{2}} & -\dfrac{1}{\sqrt{2}} \\ \dfrac{1}{\sqrt{2}} & \dfrac{1}{\sqrt{2}} \end{pmatrix}$$

FP1

Example 7

Find the matrix of a reflection in the line $y = x\sqrt{3}$.

Recognising that $\sqrt{3}$ is the tangent of 60°, use the general form:

$$\mathbf{M} = \begin{pmatrix} \cos 2\theta & \sin 2\theta \\ \sin 2\theta & -\cos 2\theta \end{pmatrix}$$

and replace the cosines and sines by the exact values of cos 120° and sin 120° to obtain:

$$\mathbf{M} = \begin{pmatrix} -\dfrac{1}{2} & \dfrac{\sqrt{3}}{2} \\ \dfrac{\sqrt{3}}{2} & \dfrac{1}{2} \end{pmatrix}$$

Exercise 4B

In Questions **1** to **6**, find the matrix of the given transformation.

1 A rotation about the origin through 30° anticlockwise.

2 A rotation about the origin through 60° clockwise.

3 A rotation about the origin through 225° anticlockwise.

4 A reflection in the line $y = \dfrac{1}{\sqrt{3}}x$.

5 A reflection in the line $y = -\dfrac{1}{\sqrt{3}}x$.

6 A reflection in the line $y = (\tan 22.5°)x$.

4.4   Identifying the geometrical transformation

In section 4.2 you saw how to find the matrix of a given transformation. In this section you will learn how to carry out the reverse process, that is finding a transformation given its matrix.

To identify the geometrical transformation you use the images of the points (1, 0) and (0, 1).

Example 8

Find the transformation represented by the matrix $\begin{pmatrix} 5 & 0 \\ 0 & 5 \end{pmatrix}$.

Using $\begin{pmatrix} 5 & 0 \\ 0 & 5 \end{pmatrix} \begin{pmatrix} 1 \\ 0 \end{pmatrix} = \begin{pmatrix} 5 \\ 0 \end{pmatrix}$, the image of $(1, 0)$ is $(5, 0)$.

Similarly, $(0, 1)$ maps onto $(0, 5)$.

Thus, the transformation is an enlargement, scale factor 5, centre at the origin.

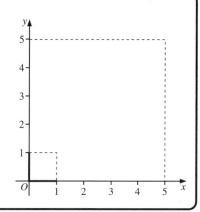

FP1

Example 9

Find the transformation represented by the matrix $\begin{pmatrix} -1 & 0 \\ 0 & -1 \end{pmatrix}$.

You have: $\begin{pmatrix} -1 & 0 \\ 0 & -1 \end{pmatrix} \begin{pmatrix} 1 \\ 0 \end{pmatrix} = \begin{pmatrix} -1 \\ 0 \end{pmatrix}$

and $\begin{pmatrix} -1 & 0 \\ 0 & -1 \end{pmatrix} \begin{pmatrix} 0 \\ 1 \end{pmatrix} = \begin{pmatrix} 0 \\ -1 \end{pmatrix}$

The transformation is a rotation of 180° about the origin, O.

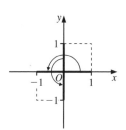

Example 10

Find the transformation represented by the matrix $\begin{pmatrix} \dfrac{1}{2} & -\dfrac{\sqrt{3}}{2} \\ \dfrac{\sqrt{3}}{2} & \dfrac{1}{2} \end{pmatrix}$.

The image of $(1, 0)$ is: $\left(\dfrac{1}{2}, \dfrac{\sqrt{3}}{2} \right)$

and the image of $(0, 1)$ is $\left(-\dfrac{\sqrt{3}}{2}, \dfrac{1}{2} \right)$.

The diagram shows that the transformation is a rotation about O, through 60° anticlockwise.

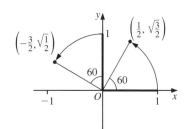

$$\cos 60° = \frac{1}{2}, \sin 60° = \frac{\sqrt{3}}{2}$$

Exercise 4C

In Questions **1** to **8**, find the transformation represented by the given matrix.

1 $\begin{pmatrix} 3 & 0 \\ 0 & 3 \end{pmatrix}$

2 $\begin{pmatrix} -2 & 0 \\ 0 & -2 \end{pmatrix}$

3 $\begin{pmatrix} 2 & 0 \\ 0 & 1 \end{pmatrix}$

4 $\begin{pmatrix} 1 & 0 \\ 0 & 6 \end{pmatrix}$

5 $\begin{pmatrix} \dfrac{1}{2} & \dfrac{\sqrt{3}}{2} \\ -\dfrac{\sqrt{3}}{2} & \dfrac{1}{2} \end{pmatrix}$

6 $\begin{pmatrix} -1 & 0 \\ 0 & -1 \end{pmatrix}$

FP1

7 $\begin{pmatrix} 1 & 0 \\ 0 & 2 \end{pmatrix}$

8 $\begin{pmatrix} \dfrac{1}{\sqrt{2}} & \dfrac{1}{\sqrt{2}} \\ -\dfrac{1}{\sqrt{2}} & \dfrac{1}{\sqrt{2}} \end{pmatrix}$

4.4   Combining two transformations

The image of the point (x, y) under **T**, the transformation represented by matrix **M**, is $\mathbf{M}\begin{pmatrix} x \\ y \end{pmatrix}$.

> **Remember:**
> $\mathbf{M_1M}$ represents the combined transformation, where **M** is the first transformation and $\mathbf{M_1}$ is the second transformation.
> $\mathbf{MM_1}$ represents the combined transformation, where $\mathbf{M_1}$ is the first transformation and **M** is the second transformation

If the image of (x, y) under the transformation **T** is transformed by a second transformation $\mathbf{T_1}$, represented by the matrix $\mathbf{M_1}$, the image of (x, y) under the two transformations is given by $\mathbf{M_1}\left(\mathbf{M}\begin{pmatrix} x \\ y \end{pmatrix}\right)$, which is $\mathbf{M_1M}\begin{pmatrix} x \\ y \end{pmatrix}$.

By the rules of matrix multiplication that you learnt in Chapter 3 (pages 27–31), you can simplify $\mathbf{M_1M}$ to a single square matrix.

In general, $\mathbf{MM_1} \neq \mathbf{M_1M}$.

Sometimes, the same image results, whatever the order of the transformations, as Examples 11, 12 and 13 show.

Example 11

Find the matrix representing the combined transformation:

✦ an enlargement, centre O, scale factor 2;

✦ followed by a rotation, centre O, anticlockwise through 90°.

First, find the images of the points $(1, 0)$ and $(0, 1)$, under the single transformations.

The matrix, **M**, of the enlargement, scale factor 2, is $\begin{pmatrix} 2 & 0 \\ 0 & 2 \end{pmatrix}$.

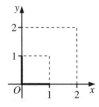

The matrix, $\mathbf{M}_1$, of the rotation, centre O, 90° anticlockwise is
$\begin{pmatrix} 0 & -1 \\ 1 & 0 \end{pmatrix}$.

The combined transformation is represented by:

$$\mathbf{M}_1\mathbf{M} = \begin{pmatrix} 0 & -1 \\ 1 & 0 \end{pmatrix}\begin{pmatrix} 2 & 0 \\ 0 & 2 \end{pmatrix}$$

$$= \begin{pmatrix} 0 & -2 \\ 2 & 0 \end{pmatrix}$$

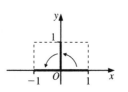

> **Make sure** you multiply the matrices in the **order $\mathbf{M}_1\mathbf{M}$**.

Notice that if you transform the point $(1, 0)$ under the two transformations in the required order, you obtain $(2, 0)$ after the first transformation, and this point $(2, 0)$ maps onto $(0, 2)$ after the second transformation.

Similarly, the point $(0, 1)$ maps onto $(-2, 0)$.

As expected, these two points are the first and second columns respectively of the final combined matrix.

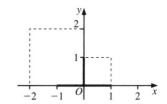

FP1

Example 12

Find the matrix representing the combined transformation:

- ✦ an enlargement, centre O, scale factor k;
- ✦ followed by a rotation about O, anticlockwise through $\theta°$.

The two matrices are:
$$\mathbf{M} = \begin{pmatrix} k & 0 \\ 0 & k \end{pmatrix} \text{ and } \mathbf{M}_1 = \begin{pmatrix} \cos\theta & -\sin\theta \\ \sin\theta & \cos\theta \end{pmatrix}$$

The combined transformation is represented by:

$$\mathbf{M}_1\mathbf{M} = \begin{pmatrix} \cos\theta & -\sin\theta \\ \sin\theta & \cos\theta \end{pmatrix}\begin{pmatrix} k & 0 \\ 0 & k \end{pmatrix}$$

$$= \begin{pmatrix} k\cos\theta & -k\sin\theta \\ k\sin\theta & k\cos\theta \end{pmatrix}$$

> **Remember**
> This matrix is given in the AQA formulae booklet.

> **Note the order**
> The enlargement (**M**) is the first transformation and the rotation (**M_1**) is the second. So multiply **M_1M** to find the matrix of the combined transformation.

In this case, you get the same matrix if you multiply $\mathbf{M}\mathbf{M}_1$. In other words, the two transformations can be done *in any order*.

You can generalise the example for any combination of a rotation and an enlargement.

In general, the matrix $\begin{pmatrix} a & -b \\ b & a \end{pmatrix}$ represents a combination of a rotation and an enlargement.

The combined transformation of a rotation about O and an enlargement, centre O, is not dependent on the order in which the transformations are carried out. A diagram shows this clearly.

a)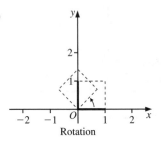
Rotation   then

b)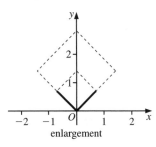
enlargement

c)
Enlargement   then

d)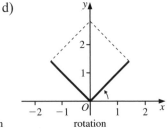
rotation

Example 13

Find the matrix representing the combined transformation:
a reflection in $y = x$ and an enlargement centre O, scale factor k.

··

The two matrices are:

$$\mathbf{M} = \begin{pmatrix} 0 & 1 \\ 1 & 0 \end{pmatrix} \text{ and } \mathbf{M_1} = \begin{pmatrix} k & 0 \\ 0 & k \end{pmatrix}.$$

a)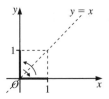
Reflection   then

b)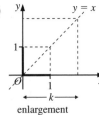
enlargement

The combined transformation is represented by:

$$\mathbf{M_1 M} = \begin{pmatrix} k & 0 \\ 0 & k \end{pmatrix}\begin{pmatrix} 0 & 1 \\ 1 & 0 \end{pmatrix}$$

$$= \begin{pmatrix} 0 & k \\ k & 0 \end{pmatrix}$$

c)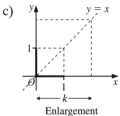
Enlargement   then

d)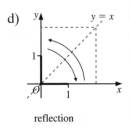
reflection

Example 14

Find the matrix representing the combined transformation:

✦ a reflection in the line $y = (\tan \theta)x$,

✦ followed by an enlargement, centre O, scale factor k.

....

The two matrices are:

$$\mathbf{M} = \begin{pmatrix} \cos 2\theta & \sin 2\theta \\ \sin 2\theta & -\cos 2\theta \end{pmatrix} \text{ and } \mathbf{M}_1 = \begin{pmatrix} k & 0 \\ 0 & k \end{pmatrix}$$

The combined transformation is represented by:

$$\mathbf{M}_1\mathbf{M} = \begin{pmatrix} k & 0 \\ 0 & k \end{pmatrix}\begin{pmatrix} \cos 2\theta & \sin 2\theta \\ \sin 2\theta & -\cos 2\theta \end{pmatrix}$$

$$= \begin{pmatrix} k \cos 2\theta & k \sin 2\theta \\ k \sin 2\theta & -k \cos 2\theta \end{pmatrix}$$

> See page 39 for the matrix of a reflection in the line $y = (\tan \theta)x$.

FP1

You can generalise the example for any combination of a reflection and an enlargement.

> In general, the matrix $\begin{pmatrix} a & b \\ b & -a \end{pmatrix}$ represents a combination of a reflection and an enlargement.

Again, these two transformations can be done in any order to produce the same image.

Example 15

Find the matrix representing the combined transformation:

✦ a reflection in the line $y = x$,

✦ followed by a stretch scale factor 3 parallel to the x-axis.

....

The two matrices are

$$\mathbf{M} = \begin{pmatrix} 0 & 1 \\ 1 & 0 \end{pmatrix} \text{ and } \mathbf{M}_1 = \begin{pmatrix} 3 & 0 \\ 0 & 1 \end{pmatrix}$$

The combined transformation is represented by

$$\mathbf{M}_1\mathbf{M} = \begin{pmatrix} 3 & 0 \\ 0 & 1 \end{pmatrix}\begin{pmatrix} 0 & 1 \\ 1 & 0 \end{pmatrix}.$$

$$= \begin{pmatrix} 0 & 3 \\ 1 & 0 \end{pmatrix}$$

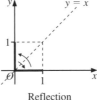

Reflection

then

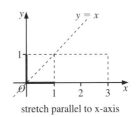

stretch parallel to x-axis

In this case, if the stretch is the first transformation and the reflection the second transformation, the combined transformation is:

$$\mathbf{MM}_1 = \begin{pmatrix} 0 & 1 \\ 1 & 0 \end{pmatrix}\begin{pmatrix} 3 & 0 \\ 0 & 1 \end{pmatrix} = \begin{pmatrix} 0 & 1 \\ 3 & 0 \end{pmatrix}$$

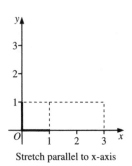

| Stretch parallel to x-axis | then | reflection |

FP1

Here, the transformations must be carried out in the correct order.

Example 16

Find the matrix representing the combined transformations:

✦   a stretch parallel to the x-axis of scale factor 2;

✦   followed by a stretch parallel to the y-axis of scale factor 5.

The two matrices are:

$$\mathbf{M} = \begin{pmatrix} 2 & 0 \\ 0 & 1 \end{pmatrix} \text{ and } \mathbf{M}_1 = \begin{pmatrix} 1 & 0 \\ 0 & 5 \end{pmatrix}$$

The combined transformation is represented by:

$$\mathbf{M}_1\mathbf{M} = \begin{pmatrix} 1 & 0 \\ 0 & 5 \end{pmatrix}\begin{pmatrix} 2 & 0 \\ 0 & 1 \end{pmatrix}$$

$$= \begin{pmatrix} 2 & 0 \\ 0 & 5 \end{pmatrix}$$

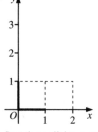

Stretch parallel to x-axis then stretch parallel to y-axis

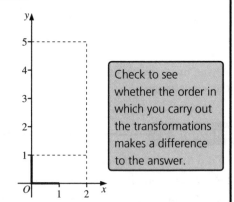

Check to see whether the order in which you carry out the transformations makes a difference to the answer.

Exercise 4D

In Questions **1** to **9**, find the matrix of the combined transformation.

1   An enlargement centre O, scale factor 3, followed by a rotation, centre O, anticlockwise through 90°.

2   An enlargement centre O, scale factor 4, followed by a rotation, centre O, clockwise through 90°.

3   A stretch parallel to the x-axis of scale factor 5, followed by a stretch parallel to the y-axis of scale factor 3.

4 A reflection in the line $y = x$, followed by a stretch parallel to the x-axis of scale factor 7.

5 A reflection in the line $y = x$, followed by a stretch parallel to the y-axis of scale factor 5.

6 A reflection in $y = x$, and an enlargement centre O, scale factor k.

7 A reflection in $y = x$, followed by a stretch parallel to the y-axis of scale factor 8, and then an enlargement, centre O, scale factor 4.

In Questions **8** to **15** describe a geometrical transformation represented by each matrix.

8 $\begin{pmatrix} -2 & 0 \\ 0 & 2 \end{pmatrix}$ **9** $\begin{pmatrix} 4 & 0 \\ 0 & -1 \end{pmatrix}$ **10** $\begin{pmatrix} 3 & 0 \\ 0 & 5 \end{pmatrix}$

FP1

11 $\begin{pmatrix} -3 & 0 \\ 0 & -3 \end{pmatrix}$ **12** $\begin{pmatrix} 3 & 0 \\ 0 & -3 \end{pmatrix}$ **13** $\begin{pmatrix} \dfrac{1}{\sqrt{2}} & \dfrac{1}{\sqrt{2}} \\ \dfrac{1}{\sqrt{2}} & -\dfrac{1}{\sqrt{2}} \end{pmatrix}$

14 $\begin{pmatrix} 1 & \sqrt{3} \\ -\sqrt{3} & 1 \end{pmatrix}$ **15** $\begin{pmatrix} \dfrac{1}{2} & \dfrac{\sqrt{3}}{2} \\ -\dfrac{\sqrt{3}}{2} & \dfrac{1}{2} \end{pmatrix}$

Summary

You should know how to ...	Check out
1 Write certain sines and cosines in surd form.	**1** Write in surd form: cos 30°, sin 45°, tan 60°
2 Represent a transformation by a matrix.	**2** Write down the matrix of each of the following transformations: a) a rotation through 120° anticlockwise about the origin b) a reflection in the y-axis c) a stretch parallel to the x-axis with scale factor 4 d) an enlargement with centre the origin and scale factor 10.

FP1

3 Identify a transformation from its matrix.

3 Give a precise geometrical description of the transformation represented by each of the following matrices:

a) $\begin{pmatrix} -1 & 0 \\ 0 & -1 \end{pmatrix}$

b) $\begin{pmatrix} 0 & 2 \\ 2 & 0 \end{pmatrix}$

c) $\begin{pmatrix} 4 & 0 \\ 0 & 3 \end{pmatrix}$.

4 Combine two transformations using their matrices.

4 The matrix $\begin{pmatrix} 0 & -1 \\ -1 & 0 \end{pmatrix}$ represents a reflection in the line $y = -x$.

The matrix $\begin{pmatrix} 1 & 0 \\ 0 & -1 \end{pmatrix}$ represents a reflection in the x-axis.

a) By multiplying these two matrices in the appropriate order, find the matrix of a reflection in the line $y = -x$ followed by a reflection in the x-axis.

b) Interpret this matrix as a single transformation.

Revision exercise 4

1 On three diagrams draw the rectangle $OABC$, where O is $(0, 0)$, $A(3, 0)$, $B(3, 2)$ and $C(0, 2)$.

For each of the following matrices calculate the coordinates of the images of O, A, B and C under the corresponding transformation, draw the image shape, and describe the transformation geometrically.

a) $\begin{pmatrix} 1 & 2 \\ 0 & 1 \end{pmatrix}$
b) $\begin{pmatrix} 1 & -\sqrt{3} \\ \sqrt{3} & 1 \end{pmatrix}$
c) $\begin{pmatrix} 1 & \sqrt{3} \\ \sqrt{3} & -1 \end{pmatrix}$.

2 The transformations $\mathbf{T}_1$ and $\mathbf{T}_2$ are represented by the matrices $\mathbf{M}_1$ and $\mathbf{M}_2$, respectively, where:

$$\mathbf{M}_1 = \begin{pmatrix} 2 & 2 \\ -2 & 2 \end{pmatrix} \text{ and } \mathbf{M}_2 = \begin{pmatrix} 3 & -3 \\ 3 & 3 \end{pmatrix}.$$

a) Describe the transformations $\mathbf{T}_1$ and $\mathbf{T}_2$ geometrically.

b) Describe geometrically the transformations represented by $\mathbf{M}_2\mathbf{M}_1$ and $\mathbf{M}_1\mathbf{M}_2$.

3 The transformations T_1 and T_2 are represented by the matrices M_1 and M_2, respectively, where:

$$M_1 = \begin{pmatrix} 2 & 0 \\ 0 & 1 \end{pmatrix} \text{ and } M_2 = \begin{pmatrix} 1 & 0 \\ 0 & 5 \end{pmatrix}.$$

a) Describe the transformations T_1 and T_2 geometrically.

b) Describe geometrically the transformations represented by $M_2 M_1$ and $M_1 M_2$.

4 The transformation T is represented by the matrix M, where:

$$M = \begin{pmatrix} 1 & 1 \\ -1 & 1 \end{pmatrix}.$$

a) Given that T is a combination of an enlargement and a rotation, find the scale factor of the enlargement and the angle of the rotation.

b) Describe the geometrical transformation represented by the matrix M^2.

c) Describe the geometrical transformations represented by the matrices M^4, M^6, and M^8.

FP1

5 The transformation T is represented by the matrix M, where:

$$M = \begin{pmatrix} \dfrac{1}{2} & -\dfrac{\sqrt{3}}{2} \\ \dfrac{\sqrt{3}}{2} & \dfrac{1}{2} \end{pmatrix}.$$

a) Give a geometrical description of T.

b) Find the smallest positive value of n for which:

$$M^n = I.$$

(AQA, 2003)

6 Two matrices A and B are such that

$$A = \begin{pmatrix} 1 & 0 \\ -1 & 1 \end{pmatrix}, B = \begin{pmatrix} 1 & 1 \\ 0 & 1 \end{pmatrix}.$$

a) Find the product BAB.

b) The transformation T is given by:

$$\begin{pmatrix} x' \\ y' \end{pmatrix} = BAB \begin{pmatrix} x \\ y \end{pmatrix}.$$

Describe fully the transformation represented by T.

(AQA, 2002)

7 The matrix **M** is $\begin{pmatrix} -\dfrac{1}{2} & -\dfrac{\sqrt{3}}{2} \\ \dfrac{\sqrt{3}}{2} & -\dfrac{1}{2} \end{pmatrix}$.

a) Find:

  i) **M**^{2};   ii) **M**^{3}.

b) The transformation **T** is given by:

$$\begin{pmatrix} x' \\ y' \end{pmatrix} = \mathbf{M} \begin{pmatrix} x \\ y \end{pmatrix}.$$

Describe fully the geometrical transformation represented by **T**.

(AQA, 2003)

8 A transformation **T** is given by:

$$\begin{pmatrix} x' \\ y' \end{pmatrix} = \frac{1}{5} \begin{pmatrix} 3 & 4 \\ -4 & 3 \end{pmatrix} \begin{pmatrix} x \\ y \end{pmatrix}.$$

a) Find the image of each of the points $A(5, 0)$ and $B(0, 5)$.

b) Describe fully the transformation represented by **T**.

(AQA, 2001)

9 A transformation **T**$_1$ is represented by the matrix:

$$\mathbf{M}_1 = \begin{pmatrix} \dfrac{\sqrt{3}}{2} & -\dfrac{1}{2} \\ \dfrac{1}{2} & \dfrac{\sqrt{3}}{2} \end{pmatrix}.$$

a) Give a geometrical description of **T**$_1$.

The transformation **T**$_2$ is a reflection in the line $y = \sqrt{3}x$.

b) Find the matrix **M**$_2$ which represents the transformation **T**$_2$.

c) i) Find the matrix representing the transformation **T**$_2$ followed by **T**$_1$.

  ii) Give a geometrical description of this combined transformation.

(AQA, 2002)

10 The transformations **T**$_1$ and **T**$_2$ are represented by the matrices **M**$_1$ and **M**$_2$, respectively, where:

$$\mathbf{M}_1 = \begin{pmatrix} 0 & 1 \\ 1 & 0 \end{pmatrix} \text{ and } \mathbf{M}_2 = \begin{pmatrix} 1 & 0 \\ 0 & -1 \end{pmatrix}.$$

a) Describe the transformations **T**$_1$ and **T**$_2$ geometrically.

b) Describe geometrically the transformations represented by **M**$_2$**M**$_1$ and **M**$_1$**M**$_2$.

c) Describe geometrically the transformation represented by **M**$_2$**M**$_1$**M**$_1$**M**$_2$**M**$_2$**M**$_1$**M**$_1$**M**$_2$.

5 Graphs of rational functions

This chapter will show you how to

- Find the equations of asymptotes to graphs of rational functions
- Find the points of intersection of these graphs with the coordinate axes
- Sketch the graphs
- Find the stationary points on these graphs
- Solve inequalities

Before you start

You should know how to ...	Check in
1 Distinguish between quadratics with two distinct linear factors, quadratics with repeated linear factors and quadratics with no real linear factors.	**1** Which of the following quadratic expressions have: a) two distinct linear factors b) a repeated linear factor c) no real linear factors? $x^2 + 4$, $x^2 - 4$, $x^2 + 4x + 4$, $2x^2 - 3x + 1$, $3x^2 + 6x + 3$
2 State when a linear expression becomes zero.	**2** For each of the following linear expressions, write down the value of x for which the expression becomes zero. a) $x - 7$ b) $7x$ c) $2x + 3$ d) $3x - 2$
3 Solve linear and quadratic equations and inequalities.	**3** Solve the following: a) $3x - 2 = 7x + 1$ b) $3x - 2 < 7x + 1$ c) $2x^2 - 3x + 1 = 0$ d) $2x^2 - 3x + 1 < 0$ e) $2x^2 - 3x + 1 > 0$

Links to Core modules

The use of the discriminant is in module C1. The solving of inequalities is explained fully in this chapter, though it would be helpful if the reader had studied the techniques included in module C1. Algebraic division, covered in C1, is not required but might be useful in a few instances. Note that, although stationary points are studied in this chapter, no knowledge of differentiation is required. Also note that you may find a graphics calculator helpful.

A rational function is an algebraic fraction in which both the numerator and denominator are polynomials. So,

$$\frac{4x-8}{x+3}, \ \frac{x^2-7}{x+1} \ \text{and} \ \frac{3x^2-2x-5}{x^2+x+1}$$

are examples of rational functions.

5.1 Rational functions with a linear numerator and denominator

When the numerator and denominator are both linear expressions, the rational function has the form:

$$y=\frac{ax+b}{cx+d}$$

where a, b, c and d are constants.

> An expression such as $ax+b$ is linear: the graph of $y=ax+b$ is a straight line.

FP1

Consider the function $y=\dfrac{4x-8}{x+3}$. To understand the function, it is useful to sketch its graph.

The function $y=\dfrac{4x-8}{x+3}$ has a linear numerator and denominator.
Notice that the graph cannot contain a point for which $x=-3$, since you cannot divide by zero.

As x approaches the value -3, the numerator approaches the value -20, but the denominator approaches 0. This means that y will have a very large positive or a very large negative value and the curve will approach the line $x=-3$ without ever touching it.

> Since $(4 \times {}^-3)-8=-20$

> Since $-3+3=0$.

The line $x=-3$ is an **asymptote** to the curve $y=\dfrac{4x-8}{x+3}$.

> When you divide by a very small number, the result is a very large number.

✦ To investigate what happens to the value of the function when the value of x becomes very large, divide both the numerator and denominator by x.

$$y=\frac{(4x-8)\div x}{(x+3)\div x}$$

$$=\frac{4-\dfrac{8}{x}}{1+\dfrac{3}{x}}$$

As x gets very large, both $-\dfrac{8}{x}$ and $\dfrac{3}{x}$ approach the value zero.

You write this as:

$$\text{When } x \to \pm\infty, \ -\frac{8}{x}\to 0 \text{ and } \frac{3}{x}\to 0.$$

> When you divide by a very large number, the result is a very small number.

> The arrow means 'approaches'.

So, $y=\dfrac{4x-8}{x+3}\equiv\dfrac{4-\dfrac{8}{x}}{1+\dfrac{3}{x}}\to\dfrac{4-0}{1-0}=4$, as $x\to\pm\infty$.

As $x\to\pm\infty$, the curve approaches the line $y=4$, so this is another asymptote.

> y will never actually be equal to 4.

✦ $x = -3$ is a **vertical asymptote**, as it is parallel to the y-axis, and
    $y = 4$ is a **horizontal asymptote**, as it is parallel to the x-axis.

✦ You also need to find where the graph of $y = \dfrac{4x - 8}{x + 3}$ crosses the
    x- and y-axes.

   You do this by substituting $x = 0$ and $y = 0$ in the function.

   When $x = 0$: $y = -\dfrac{8}{3}$

   When $y = 0$: $x = 2$

So, now you have four pieces of information about the graph of
$y = \dfrac{4x - 8}{x + 3}$.

✦ $x = -3$ is an asymptote.

✦ $y = 4$ is an asymptote.

✦ When $x = 0, y = -\dfrac{8}{3}$.

✦ When $y = 0, x = 2$.

Put this information on a diagram.

✦ First, draw the asymptotes $x = -3$
    and $y = 4$, using dashed lines.

✦ Next, mark the points $(2, 0)$ and
    $\left(0, -\dfrac{8}{3}\right)$, where the curve crosses
    the axes.

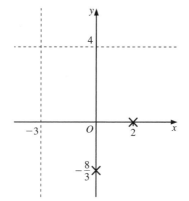

Now you need to decide how the curve
approaches the asymptotes $y = 4$ and
$x = -3$.

✦ As $x \to +\infty$, y will always be less
    than 4, so the curve approaches
    $y = 4$ from below.

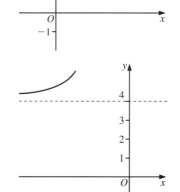

✦ As $x \to -\infty$, y will be greater
    than 4, so the curve approaches
    $y = 4$ from above.

FP1

✦ When $-3 < x < 0$, the denominator, $x + 3$, is positive but the numerator, $4x - 8$, is negative.
   So y is negative and $y \to -\infty$ as x approaches -3 from values of x greater than -3.

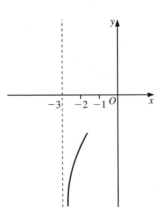

✦ When $x < -3$, the numerator and denominator are both negative. So, y is positive and $y \to +\infty$ as x approaches -3 from values of x less than -3.

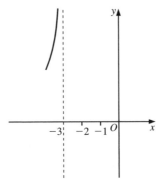

✦ Now you can complete the sketch of $y = \dfrac{4x - 8}{x + 3}$.

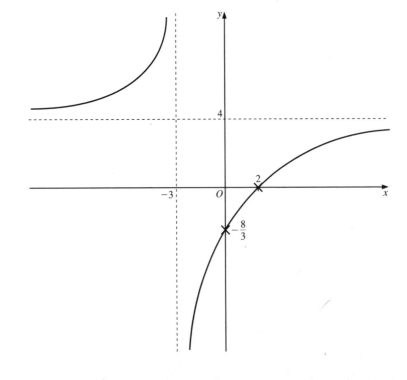

You should use a graphics calculator to verify the shape of this graph. You may find that the left-hand part of the curve does not look quite like this sketch. Sometimes, a sketch graph needs to be slightly distorted to emphasise the important features of the function.

Example 1

Sketch the curve $y = \dfrac{2x - 6}{x - 5}$.

First, find the asymptotes.

As $x \to \pm\infty$, $y \to 2$. That is, the horizontal asymptote is $y = 2$.

When $x \to 5$, $y \to \pm\infty$, so $x = 5$ is the vertical asymptote.

When x is slightly greater than 5, $y \to +\infty$.

When x is slightly less than 5, $y \to -\infty$.

Next, find where the curve crosses the axes:

When $x = 0$: $\quad y = \dfrac{-6}{-5} = \dfrac{6}{5}$

When $y = 0$: $\quad 2x - 6 = 0 \Rightarrow x = 3$

Hence, the graph of $y = \dfrac{2x - 6}{x - 5}$ looks like this:

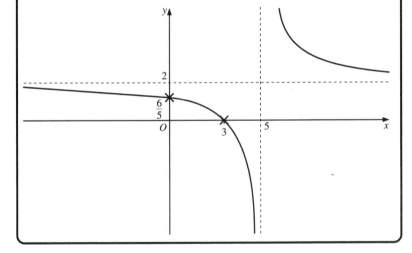

> You should use a graphics calculator or graphical software to verify the shape of this graph.

FP1

Exercise 5A

Sketch the graphs of these equations.

1 $\quad y = \dfrac{x + 1}{x - 1}$

2 $\quad y = \dfrac{x + 3}{x - 7}$

3 $\quad y = \dfrac{x - 2}{2x - 5}$

4 $\quad y = \dfrac{2x + 1}{x - 3}$

5 $\quad y = \dfrac{4x - 7}{2x + 5}$

6 $\quad y = \dfrac{3 - x}{x + 1}$

7 $\quad y = \dfrac{5 - 3x}{4 - x}$

5.2 Two distinct linear factors in the denominator

Examples of rational functions with two linear factors in the denominator are:

$$\frac{7}{(x-1)(x-2)} \qquad \frac{2x-5}{(x+1)(x-2)} \qquad \frac{5x^2+2x-1}{(x-3)(x+2)}$$

> $(x-1)$ is a linear factor.

The graphs of these functions will always have two vertical asymptotes. When you sketch such a graph, you should use a graphics calculator or graphical software to check your sketch. You can also use a graphics calculator to zoom in on the graph and explore it further.

Quadratic numerator

FP1

If the numerator is a quadratic expression and the denominator is a quadratic expression with two linear factors, for example the function:

$$y = \frac{(x-3)(2x-5)}{(x+1)(x+2)}$$

✦ there are always two vertical asymptotes;
✦ the curve will usually cross the horizontal asymptote.

> One asymptote for each factor.

To sketch the graph, you need to find four pieces of information:

✦ The horizontal asymptote.
✦ The vertical asymptotes.
✦ The points where the curve crosses the x- and y-axes.
✦ The point where the curve crosses the horizontal (y) asymptote.

Example 2

Sketch the graph of the function $y = \dfrac{(x-3)(2x-5)}{(x+1)(x+2)}$.

....

✦ To find the horizontal asymptote of $y = \dfrac{(x-3)(2x-5)}{(x+1)(x+2)}$, express the function as:

$$y = \frac{\left(1-\dfrac{3}{x}\right)\left(2-\dfrac{5}{x}\right)}{\left(1+\dfrac{1}{x}\right)\left(1+\dfrac{2}{x}\right)}$$

> Divide the numerator and denominator by x^2.

As $x \to \pm\infty$, $\dfrac{3}{x}, \dfrac{5}{x}, \dfrac{1}{x}$ and $\dfrac{2}{x} \to 0$, and $y \to 2$. Therefore, the horizontal asymptote is $y = 2$.

> $y \to \dfrac{1 \times 2}{1 \times 1} = 2$

♦ To find the vertical asymptotes, equate the denominator to zero, which gives:

$$(x + 1)(x + 2) = 0$$

Hence, the vertical asymptotes are $x = -1$ and $x = -2$.

♦ To find where the curve crosses the axes, substitute $x = 0$ and $y = 0$ in the equation.

When $x = 0$: $y = \dfrac{15}{2}$

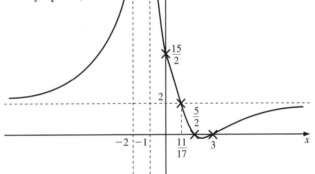

When $y = 0$: $(x - 3)(2x - 5) = 0$ and $x = 3$ or $x = \dfrac{5}{2}$

The curve crosses the y-axis at $\left(0, \dfrac{15}{2}\right)$ and the x-axis at $\left(\dfrac{5}{2}, 0\right)$ and $(3, 0)$.

♦ To find where the curve crosses the horizontal asymptote, substitute $y = 2$.

$$2 = \frac{(x - 3)(2x - 5)}{(x + 1)(x + 2)}$$

$$2(x^2 + 3x + 2) = 2x^2 - 11x + 15$$

$$6x + 4 = -11x + 15$$

$$x = \frac{11}{17}$$

The curve crosses the horizontal asymptote at $\left(\dfrac{11}{17}, 2\right)$.

To sketch the curve, insert all four points, as well as the three asymptotes.

FP1

Use a graphics calculator to check your graph.

Exercise 5B

Sketch the graphs of these functions.

1 $y = \dfrac{(x - 3)(x - 1)}{(x + 2)(x - 2)}$ **2** $y = \dfrac{(2x - 1)(x + 4)}{(x - 1)(x - 2)}$

3 $y = \dfrac{(x + 4)(x - 5)}{(x - 2)(x - 3)}$ **4** $y = \dfrac{(x + 1)(2x + 5)}{(x + 2)(x - 5)}$

5 $y = \dfrac{2x^2 + 3x - 5}{x^2 - x - 2}$ **6** $y = \dfrac{3x^2 + 4x + 4}{x^2 - 2x - 3}$

Linear numerator

> If the numerator is a linear expression and the denominator has
> two linear factors, the horizontal asymptote is **always** $y = 0$.

Example 3

Sketch the curve $y = \dfrac{2x - 9}{3x^2 - 11x + 6}$.

....

FP1

✦ To find the horizontal asymptote of $y = \dfrac{2x - 9}{3x^2 - 11x + 6}$,
express the function as

$$y = \frac{\dfrac{2}{x} - \dfrac{9}{x^2}}{3 - \dfrac{11}{x} + \dfrac{6}{x^2}}$$

> Divide the numerator and
> denominator by x^2.

> $y \to \dfrac{0 - 0}{3 - 0 + 0} = 0$

As $x \to \pm\infty$, $y \to 0$. Therefore, $y = 0$ is the horizontal asymptote.

> $y = 0$ is **always** the horizontal
> asymptote for curves of the form
> $$y = \frac{ax + b}{cx^2 + dx + e}$$

To find the vertical asymptotes, factorise the denominator.

$$y = \frac{2x - 9}{(3x - 2)(x - 3)}$$

So, when the denominator is zero, $(3x - 2)(x - 3) = 0$. Hence,

the vertical asymptotes are $x = 3$ and $x = \dfrac{2}{3}$.

To find where the curve crosses
the horizontal asymptote, in this
case the x-axis, substitute $y = 0$ in
the equation.

When $y = 0$, $2x - 9 = 0$.

Hence, $x = 4\frac{1}{2}$. So, the curve crosses the
x-axis at $(4\frac{1}{2}, 0)$.

✦ To find where the curve
crosses the y-axis,
substitute $x = 0$ in the
equation.

When $x = 0$, $y = -\dfrac{9}{6} = -1\frac{1}{2}$.

Hence, the curve crosses the y-axis at
$(0, -1\frac{1}{2})$.

Mark the asymptotes and the
two points, and sketch the curve.
Check with a graphics calculator.

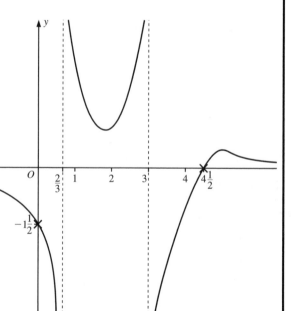

Exercise 5C

Sketch the graphs of these functions.

1 $y = \dfrac{2x + 5}{(x + 1)(x - 3)}$ **2** $y = \dfrac{3x - 6}{(x - 2)(x - 1)}$

3 $y = \dfrac{4x}{(x - 2)(x - 3)}$ **4** $y = \dfrac{x + 5}{x^2 + 7x - 8}$

5.3 Rational functions with a repeated factor in the denominator

FP1

> If the numerator is a quadratic expression and the denominator is a quadratic expression with equal factors — for example, the function $y = \dfrac{(x - 3)(x + 3)}{(x - 2)^2}$ — there is **only one** vertical asymptote.

For the function $y = \dfrac{(x - 3)(x + 3)}{(x - 2)^2}$, as x approaches 2, the denominator approaches zero, so there is a vertical asymptote, $x = 2$.

Notice that $(x - 2)^2$ is always positive.

As x approaches 2, the numerator approaches $-1 \times 5 = -5$, and is it **always** negative, whether x is just below or just above 2.

Thus, the fraction does **not** change sign from the left-hand side of the asymptote to the right-hand side.

The curve will approach the vertical asymptote through negative values of y on both sides of the asymptote.

Apart from that, the sketching process is the same as before.

Example 4

Sketch the graph of the function $y = \dfrac{(x - 3)(x + 3)}{(x - 2)^2}$.

$$y = \frac{(x - 3)(x + 3)}{(x - 2)^2} = \frac{\left(1 - \dfrac{3}{x}\right)\left(1 + \dfrac{3}{x}\right)}{\left(1 - \dfrac{2}{x}\right)^2}$$

> Divide the numerator and denominator by x^2.

As $x \to \infty$, $y \to 1$. Hence, the horizontal asymptote is $y = 1$.

The vertical asymptote is $x = 2$.

The curve crosses the axes at $\left(0, -\frac{9}{4}\right)$, $(3, 0)$ and $(-3, 0)$.

The curve crosses the horizontal asymptote $y = 1$ when:

$$1 = \frac{x^2 - 9}{x^2 - 4x + 4}$$

$$x^2 - 4x + 4 = x^2 - 9$$

$$13 = 4x$$

$$x = \frac{13}{4}$$

Now sketch the curve.

> When $y = 0$, $(x - 3)(x + 3) = 0$ and $x = +3$ or -3.

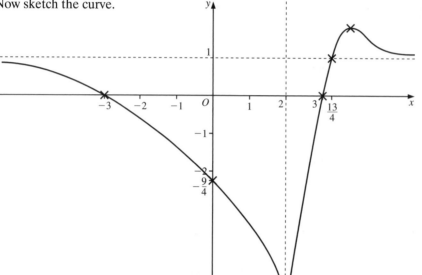

> The curve crosses the horizontal asymptote when $x = \frac{13}{4}$. Since it must approach its asymptote $y = 1$ as $x \to \infty$, it must have a maximum point for a value of x greater than $\frac{13}{4}$. The two branches of the curve near the vertical asymptote, $x = 2$, both approach $-\infty$.

Example 5

Sketch the curve $y = \dfrac{(x - 1)(3x + 2)}{(x + 1)^2}$.

....

$$y = \frac{(x - 1)(3x + 2)}{(x + 1)^2} = \frac{\left(1 - \dfrac{1}{x}\right)\left(3 + \dfrac{2}{x}\right)}{\left(1 + \dfrac{1}{x}\right)^2}$$

As $x \to \infty$, $y \to 3$. Hence, the horizontal asymptote is $y = 3$.

The vertical asymptote is $x = -1$. As x approaches -1, the numerator approaches $(-2)(-1) = 2$, which is positive. The denominator consists of a squared factor, so its values are always positive.

As $x \to -1$, $y \to \infty$ and the curve approaches the horizontal asymptote on both sides through positive values of y.

The curve crosses the axes at $x = 0, y = -2$, and at $y = 0, x = 1$ or $-\dfrac{2}{3}$.

When $y = 0$, $(x - 1)(3x + 2) = 0$ and $x = 1$ or $-\frac{2}{3}$.

The curve crosses the horizontal asymptote when $y = 3$, which gives:

$$3 = \frac{3x^2 - x - 2}{x^2 + 2x + 1}$$

$$3(x^2 + 2x + 1) = 3x^2 - x - 2$$

$$6x + 3 = -x - 2$$

$$x = -\frac{5}{7}$$

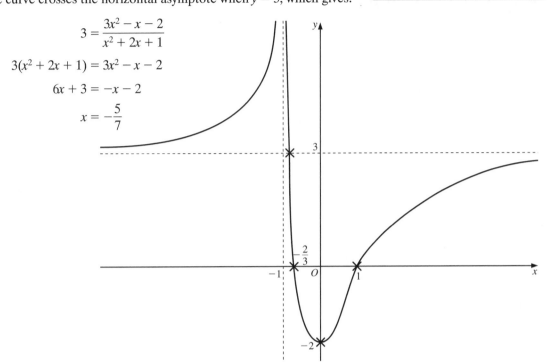

Exercise 5D

Sketch the graphs of these functions.

1 $y = \dfrac{x + 1}{(x - 1)^2}$

2 $y = \dfrac{(x - 1)(x + 1)}{(x - 3)^2}$

3 $y = \dfrac{(x + 1)(x - 2)}{(x - 1)^2}$

4 $y = \dfrac{(x + 3)(x + 4)}{(x - 2)^2}$

5 $y = \dfrac{(x + 7)(x - 5)}{(x - 1)^2}$

5.4 Rational functions with an irreducible quadratic in the denominator

Not all curves with equations of the form $y = \dfrac{x^2 + ax + b}{x^2 + cx + d}$ have vertical asymptotes.

An irreducible quadratic is one that has no real roots.

If the equation $x^2 + cx + d = 0$ does not have real solutions, then the curve will **not** have a vertical asymptote.

Example 6

Sketch the curve $y = \dfrac{x^2 + 2x - 3}{x^2 + 2x + 6}$.

◆ To find the horizontal asymptote of $y = \dfrac{x^2 + 2x - 3}{x^2 + 2x + 6}$, express the function as:

$$y = \frac{1 + \dfrac{2}{x} + \dfrac{3}{x^2}}{1 + \dfrac{2}{x} + \dfrac{6}{x^2}}$$

As $x \to \infty$, $y \to 1$. Therefore, $y = 1$ is the horizontal asymptote.

FP1

◆ Since the denominator, $x^2 + 2x + 6$, has a discriminant $4 - 24 = -20 < 0$, the curve does not have a vertical asymptote.

> The discriminant $b^2 - 4ac$ is covered in the C1 module.

◆ When $y = 0$, $x^2 + 2x - 3 = 0$
$$(x + 3)(x - 1) = 0$$
$$x = -3 \quad \text{and} \quad 1$$

Hence, the curve crosses the x-axis at $(-3, 0)$ and $(1, 0)$.

When $x = 0$, $y = -\frac{1}{2}$. Hence, the curve crosses the y-axis at $(0, -\frac{1}{2})$.

◆ The curve crosses the horizontal asymptote $y = 1$ when

$$1 = \frac{x^2 + 2x - 3}{x^2 + 2x + 6}$$

That is, $x^2 + 2x + 6 = x^2 + 2x - 3$, which is impossible.

Therefore, the curve never crosses its horizontal asymptote.

Now sketch the curve.

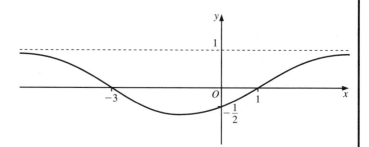

Exercise 5E

Sketch the graphs of these functions

1 $y = \dfrac{x^2}{x^2 + x + 1}$

2 $y = \dfrac{x^2 + 1}{x^2 - x + 2}$

3 $y = \dfrac{x^2 + 3x - 4}{x^2 + 3x + 7}$

4 $y = \dfrac{x^2 - 2x + 3}{x^2 + 4x + 6}$

5.5 Stationary points on the graphs of rational functions

Notice that several of the rational functions you have sketched have one or more stationary points. These stationary points can be found without using calculus.

> Maximum and minimum points are stationary points.

The curve $y = \dfrac{x^2 + 2x - 3}{x^2 + 2x + 6}$, sketched in Example 6, has a minimum point.

To find the minimum point consider a line $y = k$ intersecting the curve $y = \dfrac{x^2 + 2x - 3}{x^2 + 2x + 6}$.

At the points of intersection:

$$k = \frac{x^2 + 2x - 3}{x^2 + 2x + 6}$$

Cross-multiplying gives:

$$kx^2 + 2kx + 6k = x^2 + 2x - 3$$
$$(k - 1)x^2 + (2k - 2)x + 6k + 3 = 0 \qquad (*)$$

For equal roots:

> For the roots to be equal,
> $$b^2 - 4ac = 0$$
> For this equation, $b^2 = (2k - 2)^2$ and $4ac = 4(k - 1)(6k + 3)$.

$$(2k - 2)^2 - 4(k - 1)(6k + 3) = 0$$
$$4k^2 - 8k + 4 - 4(6k^2 - 3k - 3) = 0$$
$$-20k^2 + 4k + 16 = 0$$
$$5k^2 - k - 4 = 0$$
$$(5k^2 + 4)(k - 1) = 0$$
$$k = -\frac{4}{5} \quad \text{or} \quad k = 1$$

There is no point on the curve for which $y = 1$.

Therefore, the minimum value of y is $-\dfrac{4}{5}$.

To find the x-value at the minimum point, substitute $k = -\dfrac{4}{5}$ into the equation (*), giving:

$$-\frac{9}{5}x^2 - \frac{18}{5}x - \frac{9}{5} = 0$$
$$x^2 + 2x + 1 = 0$$
$$(x + 1)^2 = 0$$
$$x = -1$$

The minimum point has coordinates $\left(-1, -\dfrac{4}{5}\right)$.

FP1

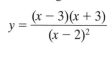

Find the stationary point on the curve

$$y = \frac{(x-3)(x+3)}{(x-2)^2}$$

> Note that this is the curve from Example 4 on page 61.

- -

◆ To find the stationary point, consider a line $y = k$ intersecting the curve

$$y = \frac{(x-3)(x+3)}{(x-2)^2}$$

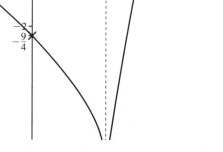

At the point of intersection:

$$k = \frac{(x-3)(x+3)}{(x-2)^2}$$

$$k(x^2 - 4x + 4) = x^2 - 9$$

$$(k-1)x^2 - 4kx + 4k + 9 = 0 \qquad (*)$$

◆ For the roots to be equal, $b^2 - 4ac = 0$.

$$(4k)^2 - 4(k-1)(4k+9) = 0$$

$$16k^2 - 4(4k^2 + 5k - 9) = 0$$

$$-20k + 36 = 0$$

$$5k - 9 = 0$$

$$k = \frac{9}{5}$$

> $b^2 = (4k)^2$ and
> $4ac = 4(k-1)(4k+9)$

Thus, the maximum value of y is $\frac{9}{5}$.

Substitute $k = \frac{9}{5}$ in the equation (*) to obtain:

$$\frac{4}{5}x^2 - \frac{36}{5}x + \frac{81}{5} = 0$$

Hence:

$$4x^2 - 36x + 81 = 0$$

$$(2x - 9)^2 = 0$$

$$x = 4\tfrac{1}{2}$$

> Again, there is a repeated value for x at a stationary point.

Hence, the stationary point is a maximum at $(4\tfrac{1}{2}, 1\tfrac{4}{5})$.

> If possible, check that your sketch agrees with the stationary point(s) you have found.

Example 8

Sketch the curve $y = \dfrac{2x^2 + 5x + 3}{4x^2 + 5x + 3}$.

Also find the values of the stationary points on the curve.

··

+ To find the horizontal asymptote of $y = \dfrac{2x^2 + 5x + 3}{4x^2 + 5x + 3}$, express the function as:

$$y = \frac{2 + \dfrac{5}{x} + \dfrac{3}{x^2}}{4 + \dfrac{5}{x} + \dfrac{3}{x^2}}$$

As $x \to \pm\infty$, $y \to \dfrac{1}{2}$. So, $y = \dfrac{1}{2}$ is the horizontal asymptote.

+ The denominator, $4x^2 + 5x + 3$, has discriminant $b^2 - 4ac = -23 < 0$, so the denominator is never zero and the curve does **not** have a vertical asymptote.

FP1

+ When $y = 0$: $\quad 2x^2 + 5x + 3 = 0$

$$(2x + 3)(x + 1) = 0$$
$$x = -1 \quad \text{or} \quad x = -\frac{3}{2}$$

The curve crosses the x-axis at $(-1, 0)$ and $\left(-\dfrac{3}{2}, 0\right)$.

When $x = 0$, $y = 1$. So, the curve crosses the y-axis at $(0, 1)$.

> When $x = 0$, $y = \frac{3}{3} = 1$

+ The curve crosses the horizontal asymptote, $y = \dfrac{1}{2}$, when:

$$\frac{1}{2} = \frac{2x^2 + 5x + 3}{4x^2 + 5x + 3}$$

which gives: $\quad 4x^2 + 5x + 3 = 4x^2 + 10x + 6$

$$x = -\frac{3}{5}$$

Now sketch the curve.

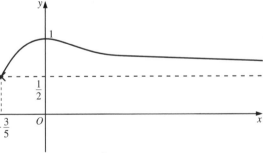

+ To find the stationary points, you need to find the values of x for which $\dfrac{2x^2 + 5x + 3}{4x^2 + 5x + 3} = k$ has equal roots.

Cross-multiplying:

$$4kx^2 + 5kx + 3k = 2x^2 + 5x + 3$$
$$(4k - 2)x^2 + (5k - 5)x + 3k - 3 = 0$$

> $b^2 - 4ac = 0$ for equal roots
> $b^2 = (5k - 5)^2$
> $4ac = 4(4k - 2)(3k - 3)$

For equal roots, the condition is:

$$(5k - 5)^2 - 4(4k - 2)(3k - 3) = 0$$
$$(23k + 1)(k - 1) = 0$$
$$k = -\tfrac{1}{23}, \quad \text{or} \quad k = 1$$

> You can use a graphics calculator to check the maximum and minimum values.

Hence, the maximum value of y is 1, and the minimum value is $-\dfrac{1}{23}$.

Exercise 5F

In Questions **1–3**, find the coordinates of the maximum and minimum points on the curve.

1 $y = \dfrac{x^2 + x - 1}{x^2 + x - 3}$ **2** $y = \dfrac{x^2 - 5}{x^2 + 2x - 11}$ **3** $y = \dfrac{x^2 + 2}{x^2 - 4x}$

In Questions **4** and **5**, sketch the curve and find the coordinates of its stationary points.

4 $y = \dfrac{2x^2 + 5x + 3}{4x^2 + 5x + 3}$ **5** $y = \dfrac{3x^2 - 9x + 7}{2x^2 - 7x + 6}$

5.6 Inequalities

FP1

You can manipulate inequalities in much the same way as equations, but with one important exception.

You can add the same number to both sides of an inequality as you do with an equation.

You can subtract the same number from both sides of an inequality as you do with an equation.

But if you **multiply or divide** both sides by a **negative** number, you must **reverse** the inequality symbol. For example:

$$4x - 1 < 5 \Rightarrow 4x < 6$$
$$3x + 2 > 11 \Rightarrow 3x > 9$$
$$\text{but} \quad 2x > 4 \Rightarrow x < -2$$

This means that you cannot solve an inequality such as $\dfrac{ax + b}{cx + d} > 2$ simply by multiplying both sides of the inequality by $cx + d$, since you do not know whether $cx + d$ is positive, in which case:

$$ax + b > 2(cx + d)$$

or negative, in which case:

$$ax + b < 2(cx + d)$$

There are two methods you can use to solve inequalities such as:

$$\frac{ax + b}{cx + d} > k$$

Method 1 Multiply both sides of the inequality by $(cx + d)^2$, which cannot be negative.

> The square of a real number is always positive or zero.

Method 2 Sketch $y = \dfrac{ax + b}{cx + d}$, solve $\dfrac{ax + b}{cx + d} = k$ and then, by comparing these two results, write down the solution to the inequality.

You should be able to use both methods, but the one which you prefer may depend on whether your algebraic skills or your graphical skills are stronger.

Example 9

Solve the inequality $\dfrac{5x - 9}{x + 3} > 2$.

···

Method 1

Multiplying by $(x + 3)^2$:

$$\frac{5x - 9}{x + 3}(x + 3)^2 > 2(x + 3)^2$$
$$(5x - 9)(x + 3) > 2(x + 3)^2$$
$$(5x - 9)(x + 3) - 2(x + 3)^2 > 0$$

Take out the factor $(x + 3)$:

$$(x + 3)[5x - 9 - 2(x + 3)] > 0$$
$$(x + 3)(3x - 15) > 0$$
$$(x + 3)(x - 5) > 0$$
$$x > 5 \quad \text{or} \quad x < -3$$

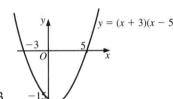

Method 2

Consider the curve $y = \dfrac{5x - 9}{x + 3}$.

The asymptotes are $x = -3$ and $y = 5$.

The curve crosses the axes at $\left(\dfrac{9}{5}, 0\right)$ and $(0, -3)$.

Now sketch the curve $y = \dfrac{5x - 9}{x + 3}$

and the line $y = 2$.

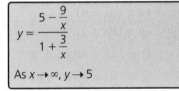

$$y = \frac{5 - \dfrac{9}{x}}{1 + \dfrac{3}{x}}$$

As $x \to \infty,\ y \to 5$

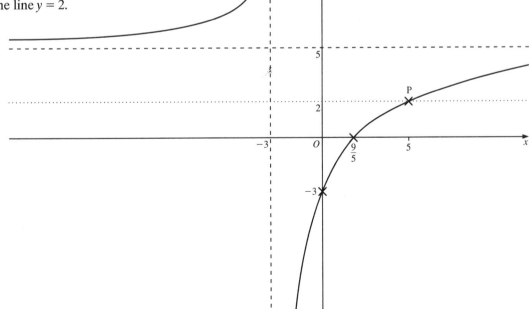

When $\dfrac{5x-9}{x+3}=2$,

$$5x-9=2(x+3)$$
$$3x=15$$
$$x=5$$

Insert the point P (5, 2) on the curve. Hence, you can see that the inequality is satisfied by the part of the graph above the dotted line $y=2$. That is, where $x>5$ or $x<-3$.

> Use the trace function on a graphics calculator to check where y is greater than 2.

Example 10

FP1

Solve the inequality $\dfrac{(x+1)(x+4)}{(x-1)(x-2)}<2$.

Method 1

> $(x-1)^2$ and $(x-2)^2$ are both positive.

Multiplying by $(x-1)^2(x-2)^2$:

$$\dfrac{(x+1)(x+4)}{(x-1)(x-2)}(x-1)^2(x-2)^2<2(x-1)^2(x-2)^2$$
$$(x+1)(x+4)(x-1)(x-2)<2(x-1)^2(x-2)^2$$
$$(x+1)(x+4)(x-1)(x-2)-2(x-1)^2(x-2)^2<0$$

Take out the factors $(x-1)$ and $(x-2)$:

$$(x-1)(x-2)[(x+1)(x+4)-2(x-1)(x-2)]<0$$
$$(x-1)(x-2)[(x^2+5x+4-2x^2+6x-4]<0$$
$$(x-1)(x-2)(-x^2+11x)<0$$
$$(x-1)(x-2)(x^2-11x)>0$$
$$(x-1)(x-2)x(x-11)>0$$

> Multiply $(-x^2+11x)$ by -1 and **reverse** the inequality.

Therefore,

$$\dfrac{(x+1)(x+4)}{(x-1)(x-2)}<2$$

when $x<0,\ 1<x<2,\ x>11$.

> When $x>11$, all factors are +ve.
> When $1<x<2$, $(x-1)$ is +ve, $(x-2)$ −ve, x +ve, $(x-11)$ −ve; product is +ve.
> When $x<0$, all factors are −ve; product is +ve.

Method 2

Consider the curve $y=\dfrac{(x+1)(x+4)}{(x-1)(x-2)}$.

The horizontal asymptote is $y=1$.

The vertical asymptotes are $x=1$ and $x=2$.

The curve crosses the axes at $(-1, 0)$ and $(-4, 0)$ and at $(0, 2)$.

The curve crosses the line $y = 2$ when:

$$\frac{(x + 1)(x + 4)}{(x - 1)(x - 2)} = 2$$

$$x^2 + 5x + 4 = 2(x^2 - 3x + 2)$$

$$0 = x^2 - 11x$$

$$x = 0 \quad \text{and} \quad 11$$

You can now sketch the curve $y = \dfrac{(x + 1)(x + 4)}{(x - 1)(x - 2)}$

and the line $y = 2$.

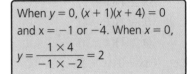

When $y = 0$, $(x + 1)(x + 4) = 0$
and $x = -1$ or -4. When $x = 0$,
$y = \dfrac{1 \times 4}{-1 \times -2} = 2$

FP1

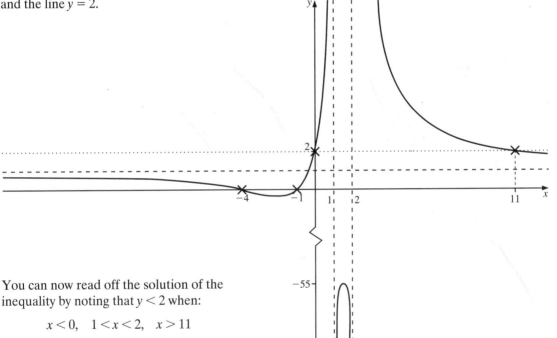

You can now read off the solution of the
inequality by noting that $y < 2$ when:

$$x < 0, \quad 1 < x < 2, \quad x > 11$$

Exercise 5G

In Questions **1** to **3**, solve each of the inequalities for x.

1 a) $\dfrac{x + 3}{x + 2} < 2$ 　　　 b) $\dfrac{x + 5}{x - 3} > 1$ 　　　 c) $\dfrac{2x - 1}{x + 3} > 3$

　 d) $\dfrac{3x + 4}{x - 5} > 2$ 　　　 e) $\dfrac{1 - 2x}{4x + 2} > 2$ 　　　 f) $\dfrac{3 + 4x}{5x - 1} > 3$

2   a) $\dfrac{(x-1)(x-2)}{(x+1)(x+2)} > 1$ b) $\dfrac{(x+2)(x-5)}{(x-3)(x-2)} > 1$

    c) $\dfrac{(x-1)(x-4)}{(x+1)(x-5)} > 2$ d) $\dfrac{(2x-1)(x-2)}{(x-3)(x+7)} > 2$

    e) $\dfrac{(x+1)(x+5)}{(x+2)(2x+3)} > 3$

3   a) $\dfrac{x^2+x-3}{x^2+x-2} > 1$ b) $\dfrac{2x^2+x-5}{2x^2+x-3} < 1$ c) $\dfrac{x^2-x-2}{x^2+3x+2} > 1$

···

Summary

FP1

You should know how to ...	Check out
1 Find the equations of asymptotes to graphs of rational functions.	**1** Find the equations of the asymptotes to the following curves: a) $y = \dfrac{3x-2}{2x+3}$ b) $y = \dfrac{3x-2}{(2x+3)(x-1)}$ c) $y = \dfrac{3x-2}{(2x+3)^2}$
2 Find the points of intersection of graphs of rational functions with the coordinate axes.	**2** Find the points of intersection of the following curves with the coordinate axes: a) $y = \dfrac{3x-2}{2x+3}$ b) $\dfrac{(3x-2)(x-1)}{2x+3}$ c) $y = \dfrac{2}{2x+3}$
3 Sketch the graphs of rational functions.	**3** Sketch the curves: a) $y = \dfrac{3x-2}{2x+3}$ b) $y = \dfrac{3x-2}{(2x+3)(x-1)}$ c) $y = \dfrac{3x-2}{(2x+3)^2}$
4 Find the stationary points on graphs of rational functions.	**4** Find the stationary points on the curve $y = \dfrac{(x+2)^2}{x^2+1}$.
5 Solve inequalities.	**5** Solve the inequality $\dfrac{x+2}{x(x-1)} > 2$.

Revision exercise 5

1 Sketch the graph of $y = \dfrac{x}{x-2}$, where $x \neq 2$.

Indicate the asymptotes and state their equations.   *(AQA, 2002)*

2 A curve has equation $y = \dfrac{x^2}{x^2 + 3x + 3}$.

a) Write down the equation of the horizontal asymptote to the curve.

b) i) Prove that, for all real values of x, y satisfies the inequality
$0 \leqslant y \leqslant 4$.
 ii) Hence, find the coordinates of the turning points on the curve.

c) Given that there are no vertical asymptotes, sketch the curve.   *(AQA, 2004)*

FP1

3 A curve C has the equation

$$y = \frac{2x+1}{x+2}, \quad x \neq -2$$

a) Express the equation of C in the form

$$y = A + \frac{B}{x+2}$$

where A and B are numbers to be found.

b) Sketch the curve C. Indicate the asymptotes and the points of intersection of the curve with the axes.   *(AQA, 2003)*

4 a) Sketch the graph of $y = \dfrac{2x-1}{x+1}$, where $x \neq -1$.

Indicate the asymptotes and the coordinates of the points of intersection of the curve with the axes.

b) Solve the inequality

$$\frac{2x+1}{x+2} < 5$$
    (AQA, 2001)

5 The graph of $y = f(x)$ is sketched below.

The asymptotes have equations $x = 1$ and $y = 4$.

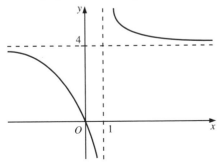

Given that $y = \dfrac{ax}{x-b}$, use the sketch to find the values of a and b.   *(AQA, 2004)*

6 Solve the inequality

$$\frac{3x}{x-2} < 4.$$

(AQA/NEAB, 2000)

7 a) Sketch the graph of $y = \dfrac{x}{(x-3)^2}$.

  b) By considering the intersection of this graph with the line $y = k$, find the coordinates of the minimum point on the curve.

  c) Solve the inequality $\dfrac{x}{(x-3)^2} < 2$.

8 a) Sketch the graph of $y = \dfrac{4x-3}{(x-1)(x-2)}$.

  b) Hence, or otherwise, solve the inequality $\dfrac{4x-3}{(x-1)(x-2)} > 0$.

FP1

9 The curve C has equation $y = \dfrac{x^2}{x^2+1}$.

  a) Show that C has no vertical asymptotes.

  b) Find the equation of the horizontal asymptote to C.

  c) By considering the intersection of C with the line $y = k$, show that the origin is the only stationary point on C.

  d) Sketch the curve C.

  e) Solve the inequality $\dfrac{x^2}{x^2+1} < \dfrac{1}{2}$.

6 Conics

This chapter will show you how to

- ✦ Recognise the standard equations of parabolas, ellipses and hyperbolas
- ✦ Sketch these curves
- ✦ Interpret the intersection of a straight line with a conic
- ✦ Translate, stretch and reflect these curves

Before you start

You should know how to ...	Check in
1 Use the discriminant of a quadratic expression to determine the nature of the roots of a quadratic equation.	**1** State, giving reasons, whether each of these equations has real distinct roots, equal roots or no real roots: a) $x^2 + 4x + 3 = 0$ b) $x^2 + 4x + 4 = 0$ c) $x^2 + 4x + 5 = 0$ d) $2x^2 - x + 10 = 0$ e) $4x^2 - 12x + 9 = 0$ f) $6x^2 + x - 3 = 0$
2 Solve quadratic equations.	**2** Solve the following quadratic equations: a) $x^2 + 4x + 3 = 0$ b) $x^2 + 4x + 4 = 0$ c) $2x^2 - 3x + 1 = 0$ d) $3x^2 = 4x - 1$
3 Transform the equation of a curve to represent a translation, a stretch or a reflection in the line $y = x$.	**3** Find the equation of the curve obtained by applying each of the following transformations to the curve $y = x^2$. (Start afresh with $y = x^2$ for each part.) a) A translation of 2 units in the positive x-direction. b) A translation of 3 units in the negative y-direction. c) A stretch in the x-direction with scale factor 2. d) A stretch in the y-direction with scale factor 4. e) A reflection in the line $y = x$.

Links to Core modules

The use of the discriminant is in module C1. Transformations of graphs are in module C2. Note that the standard formulae for conics and for the asymptotes to hyperbolas do not need to be memorised for examination purposes, nor do you need to be able to derive these formulae. Also note that a graphics calculator will help you in this chapter.

If you take a cone and cut a plane section through it in any direction, you get a curve which is a member of the class of curves called **conics** or **conic sections**.

The shape of the curve depends on the direction of the cut. That is, on the inclination (or angle), θ, of the plane section to the axis of the cone, as the diagram shows.

FP1

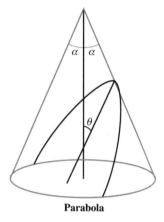

Parabola

If you cut in a direction parallel to the slant height of the cone, so that $\theta = \alpha$, you get a **parabola**.

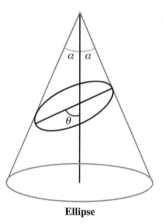

Ellipse

If you cut in a direction for which $\alpha < \theta < \dfrac{\pi}{2}$, you get an **ellipse**.

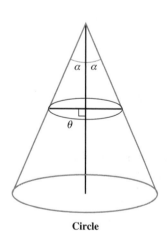

Circle

If you cut horizontally through the cone (that is, $\theta = \dfrac{\pi}{2}$), you get a **circle**.

Hyperbola

If you cut in a direction, but not through the vertex, for which $\theta < \alpha$, you get a **hyperbola** (or one branch of a hyperbola).

The study of the parabola, the ellipse and the hyperbola as sections of the same cone originated with the Greek geometer Apollonius, who flourished around 280 BC.

6.1 The parabola

The curve $y = x^2$ is a parabola.

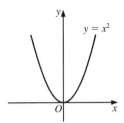

$y = x^2$

If you reflect this parabola in the line $y = x$, you get the curve $x = y^2$.

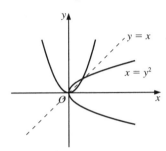

$y = x$

$x = y^2$

FP1

If the scales on the two axes are not the same, the curves will still be parabolas.

This parabola passes through the points:

$(4, 2), (1, 1), (0, 0), (1, -1), (4, -2)$

When the parabola $y^2 = x$ is stretched parallel to the x-axis, with

scale factor $\dfrac{1}{4a}$, where $a > 0$, the point $(1, 1)$ on the parabola is

transformed into the point $\left(\dfrac{1}{4a}, 1\right)$.

$\left(\dfrac{1}{4a}, 1\right)$ is on the new parabola.

If the equation of the parabola is $y^2 = kx$, you have:

$$1 = \frac{k}{4a}$$

which gives:

$$k = 4a$$

So, the new parabola has the equation

$$y^2 = 4ax$$

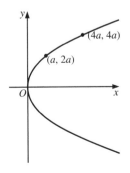

$(4a, 4a)$

$(a, 2a)$

Notice that the points $(0, 0)$, $(a, 2a)$ and $(4a, 4a)$ lie on this parabola.

The standard equation for a parabola is $y^2 = 4ax$.

Learn this standard equation.

Intersection of a line and a curve

Consider the straight line $3y = x + 18$ and the parabola $y^2 = 8x$.

To investigate the possible points of intersection of these graphs, consider the simultaneous equations $3y = x + 18$ and $y^2 = 8x$.

Eliminating x from the two equations will lead to a quadratic equation in y.

> Alternatively, eliminating y from the two equations will lead to a quadratic equation in x.

The quadratic equation may have equal roots, distinct real roots, or no real roots.

FP1

If the equation has **equal roots**, then the straight line is a tangent to the curve.

If the equation has **two distinct real roots** then the straight line is a secant to the curve.

If the equation has **no real roots** then the straight line does not intersect the curve at all.

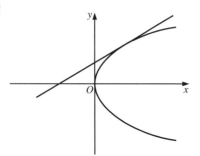

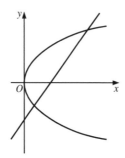

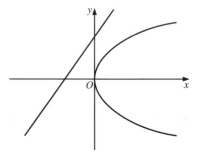

You can find the coordinates of the point of contact by solving the equation.

You can find the coordinates of the points by solving the equation.

The equation will have no real solution.

Applying these principles to the example, you can eliminate y, leading to the equation

$$\left(\frac{x + 18}{3}\right)^2 = 8x$$

which simplifies to:

$$x^2 + 36x + 324 = 72x \quad \text{or} \quad x^2 - 36x + 324 = 0$$

The discriminant of this equation is $(-36)^2 - 4(1)(324) = 0$, so the line is a tangent to the curve.

The x-coordinate of the point of contact is the repeated root of the equation (18) and the y-coordinate can be found by substituting this value of x into the equation $3y = x + 18$, giving $y = 12$.

The line touches the curve at (18, 12).

> Alternatively, you can eliminate x, leading to the equation
> $$y^2 = 8(3y - 18)$$
> or $\quad y^2 - 24y + 144 = 0$.
> The discriminant of this equation is $(-24)^2 - 4(1)(144) = 0$, so the line is a tangent to the curve. The y-coordinate of the point of contact is the repeated root of the equation (12), and the x-coordinate can be found by substituting this value of y into the equation $3y = x + 18$, giving $x = 18$.

Example 1

Sketch the curve $y^2 = 16x$.

··

The curve passes through the points:

$(0, 0), (1, 4), (1, -4), (4, 8), (4, -8), \ldots$

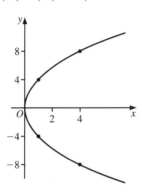

FP1

Example 2

Find the coordinates of any point or points where the line $y = x + 1$ meets the parabola $y^2 = 8x$.

··

Eliminating y:

$$(x + 1)^2 = 8x$$
$$x^2 + 2x + 1 = 8x$$
$$x^2 - 6x + 1 = 0$$

The discriminant of this equation is $(-6)^2 - 4(1)(1) = 32$, which is positive, so there are two points of intersection of the line and the curve.

The x-coordinates are:

$$\frac{6 \pm \sqrt{32}}{2} = 3 \pm \sqrt{8}$$

and the y-coordinates are found from the equation $y = x + 1$, giving the required coordinates:

$$(3 + \sqrt{8}, 4 + \sqrt{8}) \quad \text{and} \quad (3 - \sqrt{8}, 4 - \sqrt{8})$$

Exercise 6A

1 Find the coordinates of any points of intersection of the parabola $y^2 = 20x$ with the line:

a) $y = 5x$ b) $2y = x + 20$

2 Find the coordinates of any points where the parabola $y^2 = 32x$ intersects the line:

a) $y = 4x$ b) $3y = x + 72$ c) $y = 4x + 5$

3 Find the coordinates of any points where the parabola $y^2 = 60x$ intersects the line:

a) $y = 5x$ b) $y = 12x$

c) $y = 6x + 5$ d) $y = x + 15$

FP1

Transformations of a parabola

Using techniques from module C2, you should be able to deduce how the shape of a parabola is affected by certain transformations, and the effect which these transformations have on the parabola's equation.

Translation

The translation $\begin{pmatrix} 5 \\ 2 \end{pmatrix}$ moves the parabola $y^2 = 8x$ into the position shown.

Its new equation is $(y - 2)^2 = 8(x - 5)$.

Similarly the translation $\begin{pmatrix} -2 \\ 6 \end{pmatrix}$ moves the parabola $y^2 = 8x$ to the

parabola with equation $(y - 6)^2 = 8(x + 2)$.

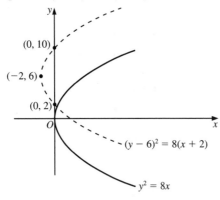

This parabola intersects the y-axis when $x = 0$, that is when:

$$(y - 6)^2 = 16$$
$$y - 6 = \pm 4$$
$$y = 10 \quad \text{or} \quad 2$$

So this parabola, after the translation $\begin{pmatrix} -2 \\ 6 \end{pmatrix}$, intersects the y-axis at the points $(0, 10)$ and $(0, 2)$.

Reflection in the line y = x

Reflecting $y^2 = 8x$ in the line $y = x$ results in the parabola shown in the diagram.

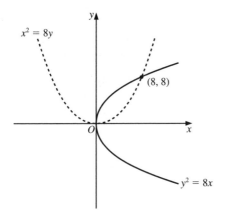

The new equation is $x^2 = 8y$.

Stretch parallel to the x-axis

A stretch, scale factor 5, parallel to the x-axis transforms the parabola $y^2 = 8x$ as shown.

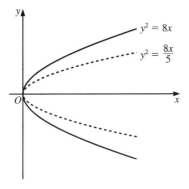

> The point (2, 4) on $y^2 = 8x$ has been transformed into the point (10, 4). Check that the point (10, 4) lies on the curve $y^2 = \dfrac{8x}{5}$.

Its equation is now $y^2 = 8\left(\dfrac{x}{5}\right)$ or $y^2 = \dfrac{8x}{5}$.

Stretch parallel to the y-axis

A stretch, scale factor 7, parallel to the y-axis transforms the parabola as shown.

Its equation is now $\left(\dfrac{y}{7}\right)^2 = 8x$ or $y^2 = 392x$.

Again, it is helpful to note that the point $(1, \sqrt{8})$ on $y^2 = 8x$ is transformed into the point $(1, 7\sqrt{8})$.

As with the transformations in Chapter 4, these transformations can be combined.

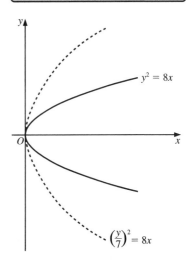

FP1

Example 3

The parabola $y^2 = 25x$ is transformed by a reflection in the line $y = x$ followed by a stretch parallel to the x-axis, scale factor 4.

Find the equation of the new parabola.

....

After the reflection in the line $y = x$, the equation of the parabola is $x^2 = 25y$.

After the stretch, this equation becomes $\left(\dfrac{x}{4}\right)^2 = 25y$ or $x^2 = 400y$.

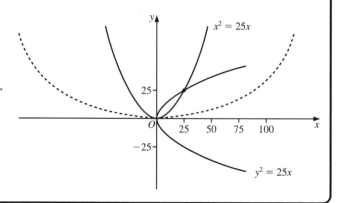

FP1

Exercise 6B

In Questions **1** to **5**, sketch the parabola before and after the transformation.

1 $y^2 = 16x$ The parabola is transformed by a reflection in the line $y = x$.

2 $y^2 = 24x$ The parabola is transformed by a translation $\begin{pmatrix} 4 \\ 3 \end{pmatrix}$.

3 $y^2 = 20x$ The parabola is transformed by a stretch parallel to the x-axis, scale factor 4.

4 $y^2 = 28x$ The parabola is transformed by a stretch parallel to the y-axis, scale factor 3.

5 $y^2 = 12x$ The parabola is transformed by a translation $\begin{pmatrix} 2 \\ -5 \end{pmatrix}$.

6 The parabola $y^2 = 16x$ is transformed by a stretch parallel to the y-axis, scale factor 2, followed by a reflection in the line $y = x$. Find the equation of the new parabola.

7 The parabola $y^2 = 36x$ is transformed by a reflection in the line $y = x$ followed by a stretch parallel to the x-axis, scale factor 7. Find the equation of the new parabola.

6.2 The ellipse

You can obtain an ellipse from a circle by applying a one-way stretch.

Consider the circle $x^2 + y^2 = 1$.

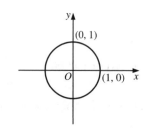

> Ellipses occur in the real world. The shadow of a circle is in the shape of an ellipse.

Applying a stretch parallel to the x-axis, with scale factor 2, will transform the circle into an ellipse.

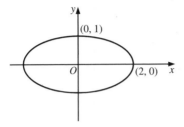

Notice that this ellipse intersects the axes at $(2, 0)$, $(-2, 0)$, $(0, 1)$ and $(0, -1)$.

The equation of the ellipse is

$$\left(\frac{x}{2}\right)^2 + y^2 = 1 \quad \text{or} \quad \frac{x^2}{4} + y^2 = 1.$$

FP1

The radius and the tangent of a circle always meet at right angles but their transformations in the ellipse do not generally meet at right angles.

To obtain the general equation of an ellipse, start with the circle $x^2 + y^2 = 1$ and apply a stretch parallel to the x-axis, scale factor a, followed by a stretch parallel to the y-axis, scale factor b.

The first stretch results in the ellipse with equation:

$$\left(\frac{x}{a}\right)^2 + y^2 = 1$$

and the second transformation results in the ellipse with equation:

$$\left(\frac{x}{a}\right)^2 + \left(\frac{y}{b}\right)^2 = 1$$

The standard equation for an ellipse is:
$$\frac{x^2}{a^2} + \frac{y^2}{b^2} = 1$$

Learn this standard equation.

To find the points of intersection of this ellipse with the axes:

When $x = 0$ $y^2 = b^2$,   so $y = \pm b$

When $y = 0$ $x^2 = a^2$,   so $x = \pm a$

Hence, the ellipse cuts the axes at $(a, 0)$, $(-a, 0)$, $(0, b)$ and $(0, -b)$.

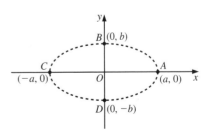

The longer axis of the ellipse, AC, is called the **major axis**, and the shorter axis, BD, is called the **minor axis**. You can see that the length of the major axis is $2a$ and that of the minor axis is $2b$.

Example 4

Find the points of intersection of the ellipse $\dfrac{x^2}{9} + \dfrac{y^2}{4} = 1$ and the line $y = 2x$.

....

Substituting $y = 2x$ into the equation $\dfrac{x^2}{9} + \dfrac{y^2}{4} = 1$ gives

$$\frac{x^2}{9} + \frac{4x^2}{4} = 1$$

This gives $10x^2 = 9$. From which:

$$x = \pm\frac{3}{\sqrt{10}} \quad \text{or} \quad \pm\frac{3\sqrt{10}}{10}$$

Using $y = 2x$ gives the points of intersection as

$$\left(\frac{3\sqrt{10}}{10}, \frac{3\sqrt{10}}{5}\right) \quad \text{and} \quad \left(-\frac{3\sqrt{10}}{10}, -\frac{3\sqrt{10}}{5}\right)$$

FP1

> $y^2 = 4x^2$

> Rationalise the denominator by multiplying numerator and denominator by $\sqrt{10}$.

Example 5

Find the points of intersection of the ellipse $\dfrac{x^2}{16} + \dfrac{y^2}{9} = 1$
and the line $3x + 4y = 12\sqrt{2}$.

....

Substituting $4y = 12\sqrt{2} - 3x$ into the equation $\dfrac{x^2}{16} + \dfrac{y^2}{9} = 1$ gives

$$\frac{x^2}{16} + \frac{(12\sqrt{2} - 3x)^2}{4^2 \times 9} = 1$$

$$9x^2 + (12\sqrt{2} - 3x)^2 = 144$$

$$9x^2 + 288 - 72\sqrt{2}x + 9x^2 = 144$$

$$18x^2 - 72\sqrt{2}x + 144 = 0$$

$$x^2 - 4\sqrt{2}x + 8 = 0$$

$$(x - 2\sqrt{2})^2 = 0$$

$$x = 2\sqrt{2} \quad \text{(Two equal roots)}$$

Substituting $x = 2\sqrt{2}$ into $3x + 4y = 12\sqrt{2}$ gives:

$$6\sqrt{2} + 4y = 12\sqrt{2}$$

$$4y = 6\sqrt{2}$$

$$y = \frac{3\sqrt{2}}{2}$$

The line $3x + 4y = 12\sqrt{2}$ touches the ellipse at $\left(2\sqrt{2}, \dfrac{3\sqrt{2}}{2}\right)$.

> Clear the fractions.

Transformations of an ellipse

In this section, you will consider the ellipse $\frac{x^2}{25} + \frac{y^2}{4} = 1$ and its image after various transformations.

Translation

The translation $\begin{pmatrix} 3 \\ 6 \end{pmatrix}$ moves the ellipse $\frac{x^2}{25} + \frac{y^2}{4} = 1$ into the position shown.

The new equation is:

$$\frac{(x-3)^2}{25} + \frac{(y-6)^2}{4} = 1.$$

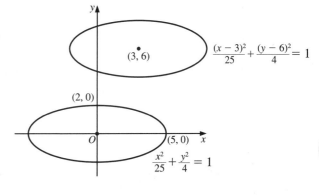

Reflection in the line y = x

Reflecting $\frac{x^2}{25} + \frac{y^2}{4} = 1$ in the line $y = x$ results in the ellipse shown.

The new equation is:

$$\frac{x^2}{4} + \frac{y^2}{25} = 1$$

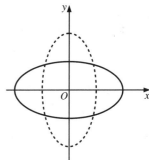

Stretch parallel to the x-axis

A stretch, scale factor 3, parallel to the x-axis transforms the ellipse

$\frac{x^2}{25} + \frac{y^2}{4} = 1$ as shown.

Its equation is:

$$\frac{\left(\frac{x}{3}\right)^2}{25} + \frac{y^2}{4} = 1 \quad \text{or} \quad \frac{x^2}{225} + \frac{y^2}{4} = 1$$

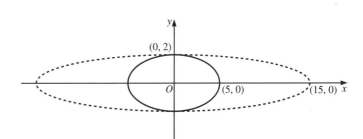

Notice that this ellipse intersects the axes at $(\pm 15, 0)$ and $(0, \pm 2)$.

FP1

Stretch parallel to the y-axis

A stretch, scale factor 7, parallel to the y-axis transforms the ellipse
$\frac{x^2}{25} + \frac{y^2}{4} = 1$ as shown.

Its equation is:

$$\frac{x^2}{25} + \frac{\left(\frac{y}{7}\right)^2}{4} = 1 \quad \text{or} \quad \frac{x^2}{25} + \frac{y^2}{196} = 1$$

Notice that this ellipse intersects the axes at $(\pm 5, 0)$ and $(0, \pm 14)$.

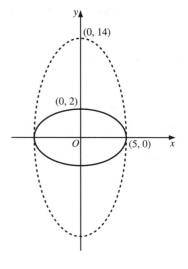

Exercise 6C

FP1

1 Sketch the ellipse $\frac{x^2}{25} + \frac{y^2}{4} = 1$, showing its points of intersection with the axes.

Find the points of intersection of this ellipse with the lines:

a) $y = 9x$ b) $y + x = 2$

2 Sketch the ellipse $\frac{x^2}{16} + \frac{y^2}{4} = 1$, showing its points of intersection with the axes.

Find the points of intersection (if any) of this ellipse with the lines:

a) $y = 4x$ b) $y = 2$ c) $4y + x = 20$

3 Find the points of intersection of the ellipse $\frac{x^2}{25} + \frac{y^2}{81} = 1$ with the lines:

a) $y = 9x$ b) $9y + x = 12$

In Questions **4** to **7** sketch the ellipse before and after the transformation given. Write down also the equation of the ellipse after the transformation.

4 $\frac{x^2}{25} + \frac{y^2}{9} = 1$ The ellipse is transformed by a reflection in the line $y = x$.

5 $\frac{x^2}{9} + \frac{y^2}{4} = 1$ The ellipse is transformed by a translation $\begin{pmatrix} 3 \\ -5 \end{pmatrix}$.

6 $\frac{x^2}{36} + \frac{y^2}{25} = 1$ The ellipse is transformed by a stretch parallel to the x-axis, scale factor 2.

7 $\dfrac{x^2}{25} + \dfrac{y^2}{36} = 1$ The ellipse is transformed by a stretch parallel to the y-axis, scale factor 4.

8 The ellipse $\dfrac{x^2}{25} + \dfrac{y^2}{4} = 1$ is transformed by a reflection in the line $y = x$ followed by a stretch parallel to the y-axis, scale factor 3.

Find the equation of the new ellipse.

9 The ellipse $\dfrac{x^2}{36} + \dfrac{y^2}{49} = 1$ is transformed by a reflection in the line $y = x$, followed by a stretch parallel to the y-axis, scale factor 5, and then by a stretch parallel to the x-axis, scale factor 2.

Find the equation of the new ellipse.

FP1

6.3 The hyperbola

A simple equation of a hyperbola is

$$y^2 = x^2 - 1 \quad \text{or} \quad x^2 - y^2 = 1$$

Note that the curve does not intersect the y-axis, but it crosses the x-axis at $(1, 0)$ and $(-1, 0)$ and has asymptotes $y = x$ and $y = -x$.

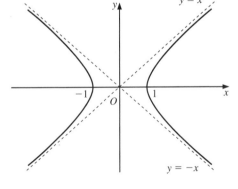

To obtain the general equation of a hyperbola, start with the equation of the hyperbola $x^2 - y^2 = 1$.

Then apply a stretch parallel to the x-axis, scale factor a,

followed by a stretch parallel to the y-axis, scale factor b.

The first transformation results in the equation:

$$\left(\frac{x}{a}\right)^2 - y^2 = 1$$

and the second transformation results in the equation:

$$\left(\frac{x}{a}\right)^2 - \left(\frac{y}{b}\right)^2 = 1$$

The standard equation for a hyperbola is:
$$\frac{x^2}{a^2} - \frac{y^2}{b^2} = 1$$

Learn this standard equation.

Note that the curve does not intersect the y-axis, but it crosses the x-axis at $(a, 0)$ and $(-a, 0)$ and has asymptotes $y = \dfrac{b}{a}x$ and $y = -\dfrac{b}{a}x$.

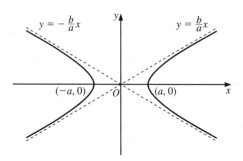

FP1

Example 6

Find the points of intersection of the hyperbola $\dfrac{x^2}{4} - \dfrac{y^2}{81} = 1$ and the line $y = 3x$.

⋯⋯⋯⋯⋯⋯⋯⋯⋯⋯⋯⋯⋯⋯⋯⋯⋯⋯⋯⋯⋯⋯⋯⋯

Substituting $y = 3x$ into the equation gives: $\dfrac{x^2}{4} - \dfrac{y^2}{81} = 1$

$$\frac{x^2}{4} - \frac{9x^2}{81} = 1$$

$$\frac{x^2}{4} - \frac{x^2}{9} = 1$$

$$5x^2 = 36$$

$$\therefore \quad x = \pm\frac{6}{\sqrt{5}} \quad \text{or} \quad \pm\frac{6\sqrt{5}}{5}$$

Substitute these values for x in the equation $y = 3x$ to find the points of intersection:

$$\left(\frac{6\sqrt{5}}{5}, \frac{18\sqrt{5}}{5}\right) \quad \text{and} \quad \left(-\frac{6\sqrt{5}}{5}, -\frac{18\sqrt{5}}{5}\right).$$

Sometimes the line does not cut the hyperbola.

Example 7

Find the points of intersection, if any, of the hyperbola $\dfrac{x^2}{36} - \dfrac{y^2}{9} = 1$ and the line $y = x$.

⋯⋯⋯⋯⋯⋯⋯⋯⋯⋯⋯⋯⋯⋯⋯⋯⋯⋯⋯⋯⋯⋯⋯⋯

Eliminating y gives:

$$\frac{x^2}{36} - \frac{x^2}{9} = 1, \text{ which simplifies to } x^2 = -12.$$

Since this equation has no real roots, the line does not intersect the hyperbola.

Transformations of a hyperbola

Translation

The translation $\begin{pmatrix} 4 \\ 2 \end{pmatrix}$ moves the hyperbola

$$\frac{x^2}{25} - \frac{y^2}{4} = 1$$

into the hyperbola as shown.

The new equation is

$$\frac{(x-4)^2}{25} - \frac{(y-2)^2}{4} = 1$$

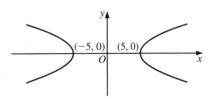

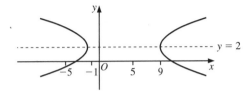

FP1

Reflection in the line y = x

Reflecting the hyperbola, $\dfrac{x^2}{25} - \dfrac{y^2}{4} = 1$, in the line $y = x$ results in the

hyperbola shown.

The new equation is:

$$\frac{y^2}{25} - \frac{x^2}{4} = 1$$

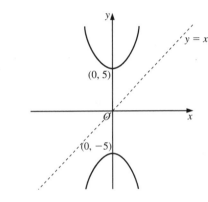

Stretch parallel to the x-axis

A stretch, scale factor 2, parallel to the x-axis transforms

the hyperbola, $\dfrac{x^2}{25} - \dfrac{y^2}{4} = 1$, as shown.

Its equation is:

$$\frac{\left(\frac{x}{2}\right)^2}{25} - \frac{y^2}{4} = 1 \quad \text{or} \quad \frac{x^2}{100} - \frac{y^2}{4} = 1$$

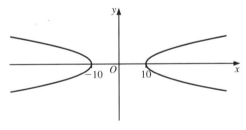

> Notice that this hyperbola intersects the x-axis at (10, 0) and (−10, 0).

Stretch parallel to the y-axis

A stretch, scale factor 3, parallel to the y-axis transforms the

hyperbola, $\dfrac{x^2}{25} - \dfrac{y^2}{4} = 1$, as shown.

Its equation is:

$$\frac{x^2}{25} - \frac{\left(\frac{y}{3}\right)^2}{4} = 1 \quad \text{or} \quad \frac{x^2}{25} - \frac{y^2}{36} = 1$$

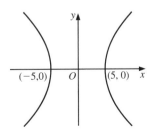

> Notice that this hyperbola intersects the x-axis at (5, 0) and (−5, 0).

Exercise 6D

1 Find the points of intersection of the hyperbola $\dfrac{x^2}{25} - \dfrac{y^2}{4} = 1$ with the lines:

    a) $9y = x$ b) $y + x = 12$

2 Find the points of intersection of the hyperbola $\dfrac{x^2}{4} - \dfrac{y^2}{9} = 1$ with the lines:

    a) $4y = x$ b) $4y + x = 2$

3 Find the points of intersection of the hyperbola $\dfrac{x^2}{16} - \dfrac{y^2}{25} = 1$ with the lines:

FP1

    a) $5y = x$ b) $4y + 5x = 1$

In Questions **4** to **7** sketch the hyperbola before and after the transformation given. Also write down the equation of the hyperbola after the transformation.

4 $\dfrac{x^2}{16} - \dfrac{y^2}{4} = 1$ The hyperbola is transformed by a reflection in the line $y = x$.

5 $\dfrac{x^2}{16} - \dfrac{y^2}{81} = 1$ The hyperbola is transformed by a translation $\begin{pmatrix} 4 \\ 3 \end{pmatrix}$.

6 $\dfrac{x^2}{9} - \dfrac{y^2}{49} = 1$ The hyperbola is transformed by a stretch parallel to the x-axis, scale factor 4.

7 $\dfrac{x^2}{16} - \dfrac{y^2}{75} = 1$ The hyperbola is transformed by a stretch parallel to the y-axis, scale factor 5.

8 The hyperbola $\dfrac{x^2}{32} - \dfrac{y^2}{25} = 1$ is transformed by a reflection in the line $y = x$, followed by a stretch parallel to the y-axis, scale factor 2. Find the equation of the new hyperbola.

9 The hyperbola $\dfrac{x^2}{32} - \dfrac{y^2}{9} = 1$ is transformed by a reflection in the line $y = x$, followed by a stretch parallel to the x-axis, scale factor 7. Find the equation of the new hyperbola.

6.4 The rectangular hyperbola

The standard equation for a hyperbola is:

$$\frac{x^2}{a^2} - \frac{y^2}{b^2} = 1$$

and the hyperbola has asymptotes:

$$y = \frac{b}{a}x \quad \text{and} \quad y = -\frac{b}{a}x$$

When $a = b$, the asymptotes of the hyperbola are $y = x$ and $y = -x$, which are perpendicular to each other. Hence, such a hyperbola (shown on the right) is called a **rectangular hyperbola**.

The equation of this rectangular hyperbola is:

$$x^2 - y^2 = a^2$$

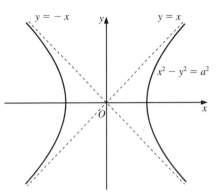

If you rotate the curve through 45° you obtain the following curve, with equation $xy = c^2$.

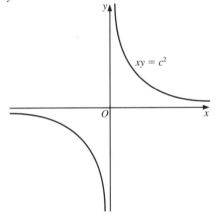

Put $b = a$ in the standard equation

$$\frac{x^2}{a^2} - \frac{y^2}{b^2} = 1 \text{ to get:}$$

$$\frac{x^2}{a^2} - \frac{y^2}{a^2} = 1$$

$$x^2 - y^2 = a^2$$

FP1

The standard equation for a rectangular hyperbola is

$$xy = c^2$$

Learn this standard equation.

Example 8

Find the point of intersection of the rectangular hyperbola $xy = 16$ and the line $y + 4x = 16$.

....

Substituting $y = 16 - 4x$ into the equation of the hyperbola $xy = 16$ gives:

$$x(16 - 4x) = 16$$
$$4x^2 - 16x + 16 = 0$$
$$(x - 2)^2 = 0$$
$$x = 2 \quad \text{(Two equal roots)}$$

Substituting into $y + 4x = 16$, you obtain $y = 8$. The line $y + 4x = 16$ touches the rectangular hyperbola $xy = 16$ at the point $(2, 8)$.

Check the point of intersection using a graphics calculator.

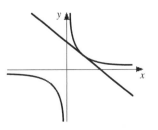

Transformations of a rectangular hyperbola

Translation

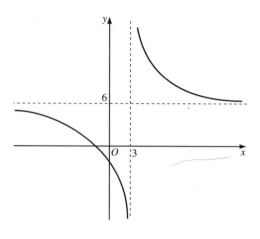

The translation $\begin{pmatrix} 3 \\ 6 \end{pmatrix}$ moves the rectangular hyperbola

$$xy = 25$$

into the rectangular hyperbola as shown.

The asymptotes are $x = 3$ and $y = 6$.

The new equation is $(x - 3)(y - 6) = 25$.

Reflection in the line $y = x$

Reflecting $xy = 25$ in the line $y = x$ results in the same rectangular hyperbola.

The equation is still $xy = 25$.

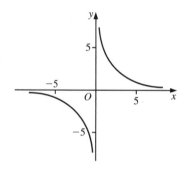

Stretch parallel to the x-axis

A stretch, scale factor 4, parallel to the x-axis transforms the rectangular hyperbola $xy = 25$ as shown.

The new equation is $\left(\dfrac{x}{4}\right)y = 25$ or $xy = 100$.

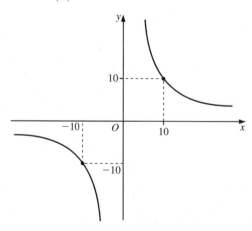

Stretch parallel to the y-axis

A stretch, scale factor 5, parallel to the y-axis transforms the rectangular hyperbola $xy = 25$ as shown.

The new equation is $x\left(\dfrac{y}{5}\right) = 25$ or $xy = 125$.

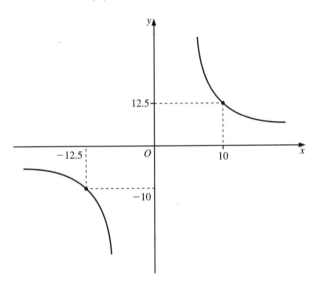

FP1

Exercise 6E

1 Find the points of intersection of the rectangular hyperbola $xy = 36$ with the lines:

 a) $y = 9x$ b) $y + x = 12$

2 Find the points of intersection of the rectangular hyperbola $xy = 25$ with the lines:

 a) $y = 4x$ b) $4y + x = 20$

 What can you deduce about the line $4y + x = 20$?

3 Find the points of intersection of the rectangular hyperbola $xy = 4$ with the lines:

 a) $y = 9x$ b) $9y + x = 12$

 What can you deduce about the line $9y + x = 12$?

In Questions **4** to **7** sketch the rectangular hyperbola before and after the transformation given. Write down also the equation of the rectangular hyperbola after the transformation.

4 $xy = 36$ The rectangular hyperbola is transformed by a reflection in the line $y = x$.

5 $xy = 49$ The rectangular hyperbola is transformed by a translation $\begin{pmatrix} 4 \\ 3 \end{pmatrix}$.

6 $xy = 81$ The rectangular hyperbola is transformed by a stretch parallel to the x-axis, scale factor 4.

7 $xy = 81$ The rectangular hyperbola is transformed by a stretch parallel to the y-axis, scale factor 5.

8 The rectangular hyperbola $xy = 81$ is transformed by a reflection in the line $y = x$, followed by a stretch parallel to the y-axis, scale factor 3.

Find the equation of the new rectangular hyperbola.

9 The rectangular hyperbola $xy = 81$ is transformed by a stretch parallel to the y-axis, scale factor 4, and then by a stretch parallel to the x-axis, scale factor 5.

Find the equation of the new rectangular hyperbola.

FP1

Summary

You should know how to ...	Check out
1 Recognise the standard equations of parabolas, ellipses and hyperbolas.	**1** Identify the following curves: a) $\dfrac{x^2}{100} - \dfrac{y^2}{25} = 1$ b) $\dfrac{x^2}{100} + \dfrac{y^2}{25} = 1$ c) $y^2 = 24x$ d) $xy = 400$
2 Sketch these curves.	**2** Sketch the curves: a) $y^2 = 16x$ b) $xy = 9$ c) $\dfrac{x^2}{16} + \dfrac{y^2}{9} = 1$ d) $\dfrac{x^2}{16} - \dfrac{y^2}{9} = 1$
3 Interpret the intersection of a straight line with a conic.	**3** Find the points of intersection, if any, of the curve $xy = 9$ with the lines: a) $y = -4x + 13$ b) $y = -4x + 12$ c) $y = -4x + 11$
4 Translate, stretch and reflect these curves.	**4** In each of the following cases give the equation of the resulting curve. a) Translate the curve $y^2 = 24x$ two units in the negative x-direction and one unit in the positive y-direction. b) Stretch the curve $xy = 400$ by a scale factor 2 in the x-direction and by a scale factor 3 in the y-direction. c) Reflect the curve $\dfrac{x^2}{100} - \dfrac{y^2}{25} = 1$ in the line $y = x$.

Revision exercise 6

1 a) Sketch the curve $xy = 16$. What is the name given to this type of curve?

b) Find the coordinates of the points where this curve intersects the line $y = x$.

c) Sketch the curve $(x - 1)(y + 1) = 16$.

2 a) Sketch the curve $y^2 = 16x$.

b) Three lines have equations $y = x + 2, y = x + 4, y = x + 6$.

Show that one of these lines intersects the curve in two distinct points, one is a tangent to the curve, and one does not intersect the curve at all.

FP1

3 a) Sketch the curve $x^2 + y^2 = 1$. What is the name given to this type of curve?

b) Apply the following transformations in succession, writing down the equation of the transformed curve at each stage.

i)   A stretch with scale factor 3 parallel to the x-axis.

ii) A stretch with scale factor 4 parallel to the y-axis.

iii) A translation with vector $\begin{pmatrix} 1 \\ 2 \end{pmatrix}$.

4 a) Sketch the hyperbola $x^2 - y^2 = 1$.

b) Apply the following transformations in succession, writing down the equation of the transformed curve at each stage.

i)   A translation with vector $\begin{pmatrix} 1 \\ 2 \end{pmatrix}$.

ii) A stretch with scale factor 3 parallel to the x-axis.

iii) A stretch with scale factor 4 parallel to the y-axis.

5 A hyperbola has equation $(x - 1)^2 - y^2 = 1$.

a) Write down the equation of the hyperbola formed by reflecting the given hyperbola in the line $y = x$, and sketch the new curve.

b) Find the equations of the two horizontal tangents to this reflected curve.

c) Find the ranges of values of k such that the line $y = k$ intersects the reflected curve in two distinct points.

6 a) Sketch the curve $y^2 = 8x$.

b) Write down the equation of the curve obtained when the curve $y^2 = 8x$ is reflected in the line $y = x$.

c) Describe a geometrical transformation that maps the curve $y^2 = 8x$ onto the curve with equation $y^2 = 8x - 16$.

(AQA, 2004)

7 a) Sketch the graph of $\dfrac{x^2}{16} + \dfrac{y^2}{49} = 1$, marking the values of the intercepts with the coordinate axes.

b) Describe a sequence of geometrical transformations that maps the graph of $\dfrac{x^2}{16} + \dfrac{y^2}{49} = 1$ onto the graph of $\dfrac{x^2}{4} + \dfrac{(y-3)^2}{49} = 1$.

(AQA, 2003)

8 A parabola has cartesian equation $y^2 = 24x$.

The line $y = mx + c$ is a tangent to P. Show that the value of mc is constant, and find its value.

(AQA, 2003)

9 a) The graph of $\dfrac{x^2}{9} - \dfrac{y^2}{4} = 1$ is sketched below.

Identify whether it is graph G_1, G_2 or G_3.

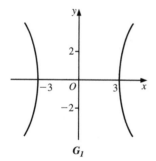
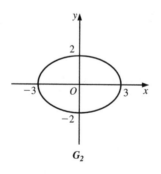
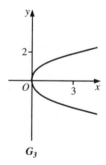

G_1

G_2

G_3

b) On separate sets of axes, sketch the graphs of:

i) $\dfrac{(x+3)^2}{9} - \dfrac{y^2}{4} = 1$

ii) $\dfrac{y^2}{9} - \dfrac{x^2}{4} = 1$

(AQA, 2002)

10 Given that the line $y = mx + c$ is a tangent to the circle $x^2 + y^2 = 1$, show that $c^2 = 1 + m^2$.

(AQA, 2002)

7 Complex numbers

This chapter will show you how to

♦ Add, subtract and multiply complex numbers
♦ Find the conjugate of a complex number
♦ Solve quadratic equations with non-real roots
♦ Equate real and imaginary parts

Before you start

FP1

You should know how to ...	Check in
1 Solve simultaneous linear equations.	**1** Solve the simultaneous equations: $7x + 3y = 1$ $\qquad$ $3x - 2y = 7$
2 Use the quadratic formula to solve quadratic equations with real roots.	**2** Use the quadratic formula to solve the equation: $$2x^2 - 3x - 4 = 0$$

> **Links to Core modules**
> Techniques from the C1 module for solving quadratic equations and simultaneous linear equations are needed in this chapter.

Mathematicians prior to the eighteenth century were used to working with a **number system**. That is, a set of objects called numbers which obeyed certain laws (the laws of arithmetic and of algebra).

However, they wanted a number system in which they would be able to solve equations like $x^2 = -1$ or $x^2 + 2x + 5 = 0$. They introduced the idea of '**imaginary numbers**', which had squares equal to negative numbers. The familiar numbers with positive or zero squares were called '**real numbers**'. They added real and imaginary numbers to get '**complex numbers**'.

One of the imaginary numbers was called i and had the property that $i^2 = -1$.

7.1 What is a complex number?

A **complex number** is a number of the form:

$$a + ib$$

where a and b are real numbers and $i^2 = -1$.

For example, $3 + 5i$ is a complex number, with $a = 3$ and $b = 5$.

In the complex number $a + ib$, the real number a is called the **real part**, and the real number b is called the **imaginary part** of the complex number.

> Note that the imaginary part is not actually an 'imaginary number'.

✦ If $a = 0$, the number is said to be **wholly imaginary**.
✦ If $b = 0$, the number is **real**.
✦ If a complex number is equal to 0, both a and b are 0.

FP1

In algebra you use letters, such as a, b, x and y, to represent unknown numbers.

Complex numbers are often denoted by the letter z, where $z = x + iy$.

7.2 Calculating with complex numbers

When you work with complex numbers, you use ordinary algebraic methods.

> Keep the real parts and the imaginary parts separate.

> If two complex numbers are equal, **their real parts are equal and their imaginary parts are equal.**
>
> This is a **necessary condition** for the equality of two complex numbers.
>
> Hence, if $a + ib = c + id$, then $a = c$ and $b = d$.

For example, if $x + iy = 2 + 3i$, then $x = 2$ and $y = 3$.

Addition and subtraction

To add two complex numbers, add the real parts and the imaginary parts separately. For example:

$$(3 + 7i) + (4 - 6i) = (3 + 4) + (7i - 6i)$$
$$= 7 + i$$

> The rule for addition is:
>
> $$(x + iy) + (u + iv) = (x + u) + i(y + v)$$
>
> and the rule for subtraction is:
>
> $$(x + iy) - (u + iv) = (x - u) + i(y - v)$$

> You do not need to learn these rules.

Example 1

Subtract $8 - 4i$ from $7 + 2i$.

$$(7 + 2i) - (8 - 4i) = (7 - 8) + (2i + 4i)$$
$$= -1 + 6i$$

Example 2

Given that $z_1 = 3 - 2i$ and $z_2 = 4 + 7i$, find:

a) $z_1 + z_2$ b) $z_1 - z_2$ c) $z_2 - z_1$

a) $z_1 + z_2 = (3 - 2i) + (4 + 7i)$
   $= 7 + 5i$

b) $z_1 - z_2 = (3 - 2i) - (4 + 7i)$
   $= -1 - 9i$

c) $z_2 - z_1 = (4 + 7i) - (3 - 2i)$
   $= 1 + 9i$

FP1

Multiplication

You can multiply complex numbers using the general algebraic method for multiplication. For example:

$$(2 + 3i)(4 - 5i) = 2(4 - 5i) + 3i(4 - 5i)$$
$$= 8 - 10i + 12i - 15i^2$$

Since $i^2 = -1$, this simplifies to:

$$8 - 10i + 12i + 15 = 23 + 2i$$

> Whenever i^2 appears in a product, replace it with -1.

The rule for multiplication is:

$$(a + ib)(c + id) = (ac - bd) + i(ad + bc)$$

> It is simpler to multiply out the numbers every time than to memorise this formula.

Example 3

Given that $z_1 = 4 + 3i$ and $z_2 = 2 - 7i$, find:

a) $z_1 z_2$ b) z_1^2

a) $z_1 z_2 = (4 + 3i)(2 - 7i)$
   $= 4(2 - 7i) + 3i(2 - 7i)$
   $= 8 - 28i + 6i - 21i^2$
   $= 8 - 28i + 6i + 21$
   $= 29 - 22i$

b) $z_1^2 = z_1 \times z_1$
   $= (4 + 3i)(4 + 3i)$
   $= 16 + 12i + 12i + 9i^2$
   $= 16 + 12i + 12i - 9$
   $= 7 + 24i$

Exercise 7A

1 Simplify these quantities.

a) i^3 b) i^4 c) i^6 d) i^9

2 Simplify these quantities.

a) $(8 + 4i) + (2 - 6i)$ b) $(-7 + 3i) + (8 - 4i)$

c) $2 - 4i + 3(-1 + 2i)$ d) $4(-2 + 5i) + 5(2 + 7i)$

e) $(8 + 3i) - (7 + 2i)$ f) $(7 + 6i) - (4 - 2i)$

g) $2(9 - 3i) - 4(2 - 6i)$ h) $3(8 + i) - 2(3 - 5i)$

FP1

3 Evaluate each of these expressions.

a) $(3 + i)(2 + 3i)$ b) $(4 - 2i)(5 + 3i)$

c) $(8 - i)(9 + 2i)$ d) $(9 - 3i)(5 - i)$

e) $i(2 - 3i)(i + 4)$ f) $(3 - 2i)(7 - 5i)$

The complex conjugate of a complex number

Consider what happens when these two complex numbers are multiplied together:

$$(2 + 3i)(2 - 3i) = 2(2 - 3i) + 3i(2 - 3i)$$
$$= 4 - 6i + 6i - 9i^2$$
$$= 4 + 9$$
$$= 13$$

The result is a real number, and $2 - 3i$ is called the **complex conjugate** (or simply the **conjugate**) of $2 + 3i$.

> In general, if z is a complex number, its complex conjugate is denoted by z^*.
>
> If $z = x + iy$, then $z^* = x - iy$.

$$zz^* = (x + iy)(x - iy) = x(x - iy) + iy(x - iy)$$
$$= x^2 - xiy + xiy - i^2y^2$$
$$= x^2 + y^2$$

> When a complex number is multiplied by its complex conjugate, the answer is a non-negative real number.

Example 4

Given that $z = -3 + 4i$, find the values of:

a) $z + z^*$ b) $z - z^*$ c) zz^*

....

a) $z + z^* = (-3 + 4i) + (-3 - 4i) = -6$

b) $z - z^* = (-3 + 4i) - (-3 - 4i) = 8i$

c) $zz^* = (-3 + 4i)(-3 - 4i) = 9 + 16 = 25$

7.3   Roots of quadratic equations

Consider the equation $x^2 + 1 = 0$.

The square of x has to be -1, which cannot be true for any real number x, but you have seen that it is true for the complex number i.

So, a solution of the equation is $x = i$.

You expect a quadratic equation to have two roots:

$$(-i)^2 = (-1)^2(i)^2 = (+1)(-1) = -1$$

Hence, $-i$ is also a solution of the equation.

So, the equation has two roots which are conjugates of each other.

Now consider $x^2 + 2x + 5 = 0$. This equation can be written as:

$$(x + 1)^2 = -4$$

Therefore, $x + 1$ has to be one of the two square roots of -4.

You know that $(2i)^2 = -4$ and that $(-2i)^2 = -4$. So:

$$x + 1 = \pm 2i$$

Therefore, the roots of the equation are $-1 \pm 2i$.

> Recall that you can use the quadratic formula:
>
> $$x = \frac{-b \pm \sqrt{b^2 - 4ac}}{2a}$$
>
> to solve quadratic equations of the form $ax^2 + bx + c = 0$.
>
> If $b^2 - 4ac$ is a negative number, you can use the equivalent form:
>
> $$x = \frac{-b \pm i\sqrt{4ac - b^2}}{2a}$$

Provided the coefficients in the equation are real but the roots are non-real, the two roots are always conjugates of each other.

They are said to form a **conjugate pair**.

FP1

> Note that $-i$, or $0 - i$, is the conjugate of i, or $0 + i$.

> Since -4 has two square roots, you can use the notation $\pm\sqrt{-4}$. However, you should not use the notation $\sqrt{-4}$ without the plus-or-minus symbol, since this would lead to the false statement:
>
> $$\sqrt{-4}\sqrt{-4} = \sqrt{16} = 4$$

> Now, $4ac - b^2$ is a positive real number that has a positive square root.

Example 5

Find the roots of $3x^2 + 4x + 6 = 0$.

Using the quadratic formula:

$$x = \frac{-4 \pm \sqrt{16 - 72}}{6}$$

$$= \frac{-4 \pm \sqrt{-56}}{6} = \frac{-4 \pm 2\sqrt{-14}}{6} = \frac{-2 \pm i\sqrt{14}}{3}$$

In chapter 1, you learnt that if α and β are the roots of the quadratic equation $ax^2 + bx + c = 0$, then:

the sum of the roots, $\alpha + \beta$, is equal to $-\dfrac{b}{a}$

FP1

and the product of the roots, $\alpha\beta$, is equal to $\dfrac{c}{a}$

These equations are still true when the roots are complex numbers.

For example, in Example 5 above the sum of the roots is:

$$\frac{-2 + i\sqrt{14}}{3} + \frac{-2 - i\sqrt{14}}{3} = -\frac{4}{3} \quad \text{which is} \quad -\frac{b}{a}$$

Example 6

The roots of the equation $3x^2 + 2x + 8 = 0$ are α and β.

Find the equation with roots α^2 and β^2.

From: $3x^2 + 2x + 8 = 0$

$$\alpha + \beta = -\frac{2}{3}$$

and $$\alpha\beta = \frac{8}{3}$$

The sum of the new roots is $\alpha^2 + \beta^2$:

$$\alpha^2 + \beta^2 = (\alpha + \beta)^2 - 2\alpha\beta$$

$$= \left(-\frac{2}{3}\right)^2 - 2 \times \frac{8}{3}$$

$$= \frac{4}{9} - \frac{16}{3}$$

$$= -\frac{44}{9}$$

The product of the new roots is:

$$\alpha^2\beta^2 = (\alpha\beta)^2$$

$$= \frac{64}{9}$$

Thus, the required equation is:

$$x^2 + \frac{44}{9}x + \frac{64}{9} = 0$$

$$9x^2 + 44x + 64 = 0$$

> Note that $\alpha^2 + \beta^2$ is negative, which could not happen if α and β were real numbers.

Example 7

The equation $x^2 - 4x + c = 0$, where c is a real number, has one root $2 - i$.

a) Write down the value of the other root of the equation.

b) Find the product of the roots.

c) Hence, write down the value of c.

a) The roots come in a conjugate pair. So, the second root is $2 + i$.

b) $(2 - i)(2 + i) = 4 + 2i - 2i - i^2$
$$= 4 + 1$$
$$= 5$$

The product of the roots is 5.

c) Since $a = 1$ and $\dfrac{c}{a} = 5$, it follows that $c = 5$.

FP1

Exercise 7B

1 Solve each of these equations.

a) $z^2 + 2z + 4 = 0$ b) $z^2 - 3z + 6 = 0$

c) $2z^2 + z + 1 = 0$ d) $4z - 3 - 2z^2 = 0$

2 Find the roots of each of these equations.

a) $x^2 + 4x + 7 = 0$ b) $x^2 + 2x + 6 = 0$

c) $2x^2 + 6x + 9 = 0$ d) $x^2 - 5x + 25 = 0$

3 The roots of the equation $x^2 + 7x + 15 = 0$ are α and β. Find the equation whose roots are 2α and 2β.

4 The roots of the equation $x^2 - 4x + 9 = 0$ are α and β. Find the equation whose roots are 3α and 3β.

5 The roots of the equation $2x^2 + 3x + 17 = 0$ are α and β. Find the equation whose roots are α^2 and β^2.

6 The roots of the equation $3x^2 - 7x + 15 = 0$ are α and β. Find the equation whose roots are α^2 and β^2.

7 The equation $2x^2 + 3x + 7 = 0$ has roots α and β. Find the equation whose roots are:

a) $2\alpha, 2\beta$ b) $\dfrac{\alpha}{3}, \dfrac{\beta}{3}$ c) α^2, β^2 d) $2\alpha + 7, 2\beta + 7$

8 The equation $3x^2 + 2x + 8 = 0$ has roots α and β. Find the equation whose roots are:

a) $4\alpha, 4\beta$ b) $\dfrac{\alpha}{2}, \dfrac{\beta}{2}$ c) α^2, β^2 d) $5\alpha + 4, 5\beta + 4$

9 The equation $x + 2 + \dfrac{3}{x} = 0$ has roots α and β.

Find the equation whose roots are 5α and 5β.

10 The roots of the quadratic equation $x^2 - 5x + 8 = 0$ are α and β.
Without solving the equation, find a quadratic equation, with
integer coefficients, whose roots are $\dfrac{1}{\alpha}$ and $\dfrac{1}{\beta}$.

FP1

7.4 Linear equations with complex coefficients

The equation $3z + z^* = 4 - 4i$ can be solved by letting $z = x + iy$.

Then z^*, the conjugate of z, is $x - iy$. Hence:

$\quad 3z + z^* = 4 - 4i$ becomes

$\quad 3(x + iy) + (x - iy) = 4 - 4i$

Equating real parts: $4x = 4 \Rightarrow x = 1$

Equating imaginary parts: $2y = -4 \Rightarrow y = -2$

So, $z = 1 - 2i$.

> **Remember**
> If two complex numbers are equal, their real parts are equal and their imaginary parts are equal.

Example 8

Find z when $4z + 2i = 8(4 - 3i)$.

$\quad\quad 4z + 2i = 8(4 - 3i)$

Substitute $x + iy$ for z:

$\quad\quad 4(x + iy) + 2i = 8(4 - 3i)$

$\quad\quad 4x + 4iy + 2i = 32 - 24i$

$\quad\quad\quad\quad 4x = 32 \quad \Rightarrow \quad x = 8$

$\quad\quad 4y + 2 = -24 \quad \Rightarrow \quad y = -\dfrac{13}{2}$

Therefore, $\quad\quad\quad\quad z = 8 - \dfrac{13}{2}i$

> Equate real and imaginary parts.

Example 9

Find z when $5z - 2z^* = 3 - 8i$.

$\quad\quad 5z - 2z^* = 3 - 8i$

Let $z = x + iy$, and then $z^* = x - iy$.

Substitute for z and z^* in the equation:

$\quad\quad 5(x + iy) - 2(x - iy) = 3 - 8i$

$\quad\quad\quad\quad 3x + 7iy = 3 - 8i$

Therefore, $3x = 3 \implies x = 1$

$\qquad 7y = -8 \implies y = -\dfrac{8}{7}$

$\qquad\qquad\qquad \therefore z = 1 - \dfrac{8}{7}\text{i}$

> Equate real and imaginary parts.

FP1

Example 10

Find z when $2z - 4\text{i} = 3\text{i}(z^* + 2) + 5 - 8\text{i}$.

$\qquad 2z - 4\text{i} = 3\text{i}(z^* + 2) + 5 - 8\text{i}$

Let $z = x + \text{i}y$, and then $z^* = x - \text{i}y$

Substitute for z and z^* in the equation:

$\qquad 2(x + \text{i}y) - 4\text{i} = 3\text{i}(x - \text{i}y + 2) + 5 - 8\text{i}$

$\qquad 2x + 2\text{i}y - 4\text{i} = 3\text{i}x + 3y + 6\text{i} + 5 - 8\text{i}$

$\qquad\qquad\qquad 2x = 3y + 5 \qquad\qquad\qquad \textbf{(1)}$

$\qquad\qquad\qquad 2y - 4 = 3x + 6 - 8$

$\qquad\qquad\qquad 2y = 3x + 2 \qquad\qquad\qquad \textbf{(2)}$

Rearranging these two simultaneous equations:

$\qquad\qquad\qquad 2x - 3y = 5 \qquad\qquad\qquad \textbf{(1)}$

$\qquad\qquad\qquad -3x + 2y = 2 \qquad\qquad\quad \textbf{(2)}$

$\qquad\qquad\qquad 4x - 6y = 10$

$\qquad\qquad\qquad -9x + 6y = 6$

$\qquad\qquad\qquad\qquad -5x = 16$

$\qquad\qquad\qquad\qquad x = -\dfrac{16}{5}$

Substitute $-\dfrac{16}{5}$ for x in equation **(2)**

$\qquad\qquad\qquad 2y = 3 \times -\dfrac{16}{5} + 2$

$\qquad\qquad\qquad y = -\dfrac{19}{5}$

So, $z = -\dfrac{16}{5} - \dfrac{19}{5}\text{i}$

> Equate real and imaginary parts.

> Multiply equation **(1)** by 2 and equation **(2)** by 3 to equate the y-coefficients.

> Add to eliminate y.

Exercise 7C

1 Find z when:

    a) $3z + 40\text{i} = 6 + 13\text{i}$ b) $2z - 5\text{i} = 3z + 1 + \text{i}$

2 Find z when:

    a) $2z - 2\text{i} = 4z^* + 6$ b) $5z - 4 = 2\text{i}z^* - 7$

3 Find z when:

a) $(3 + i)z + 2z* = 9 + 3i$ b) $(7 + 2i)z* = 5iz - 14i$

c) $(2 + 5i)z - 4z* = 2(3 - 2i)$

4 A complex number z satisfies the equation

$$iz* + 3 = (3 + 2i)z$$

where $z*$ is the complex conjugate of z.

Find z in the form $a + ib$, where a and b are real.

Summary

FP1

You should know how to ...	Check out
1 Add, subtract and multiply complex numbers.	**1** Calculate: a) $(3 + 4i) + (5 - 6i)$ b) $(2 - i) - (-3 + 2i)$ c) $(2 + i)(3 - i)$
2 Find the conjugate of a complex number.	**2** Given that $z = 3 + 4i$, calculate: a) $z + z*$ b) $z - z*$ c) $zz*$
3 Solve quadratic equations with non-real roots.	**3** Solve the equations: a) $x^2 + 9 = 0$ b) $x^2 + 4x + 5 = 0$ c) $x^2 - 6x + 16 = 0$
4 Equate real and imaginary parts	**4** Find x and y given that: $(7 + 3i)x + (3 - 2i)y = 1 + 7i$

Revision exercise 7

1 a) Show that $(1 + i)^2 = 2i$.

b) Hence find in their simplest forms the values of $(1 + i)^4$, $(1 + i)^6$ and $(1 + i)^8$.

2 Given that

$$(2 + i)(x - 2i) = y - 3i$$

where x and y are real numbers:

a) find two equations for x and y b) find the values of x and y

3 Given that

$$(1 + i)(x + yi) = 1 + 3i$$

where x and y are real numbers:

a) find two equations for x and y b) find the values of x and y

4 It is given that $z = x + iy$ and that z^* is the complex conjugate of z.

   a) Express $z + 2z^*$ in the form $p + qi$

   b) Find the value of z for which $z + 2z^* = 9 + 2i$.

5 It is given that $z = x + iy$ and that z^* is the complex conjugate of z.

   a) Express $3z - 4z^*$ in the form $p + qi$.

   b) Find the value of z for which $3z - 4z^* = 5 + 21i$.

6 Given that $z = -2 + 2\sqrt{3}i$, show that $z^2 + 4z$ is real.   *(AQA, 2002)*

7 a) Express in the form $a + ib$:

   i) $(3 + i)^2$

   ii) $(2 + 4i)(3 + i)$

   b) Verify that $3 + i$ is a root of the quadratic equation

   $z^2 - (2 + 4i)z + 8i - 6 = 0$ *(AQA, 2004)*

FP1

8 a) Find the value of the following, giving each answer in the form $a + bi$, where a and b are integers.

   i) $(2 + 3i)^2$,

   ii) $(2 + 3i)^4$.

   b) i) Given that $2 + 3i$ is a root of the equation

   $$z^4 + 40z + k = 0$$

   find the value of the real constant k.

   ii) Write down another root of the equation $z^4 + 40z + k = 0$.   *(AQA, 2004)*

9 a) The roots of the quadratic equation $x^2 + 4x + 13 = 0$ are α and β.

   Without solving the equation, find the value of

   i) $\alpha^3 + \beta^3$,

   ii) $\dfrac{\alpha}{\beta^2} + \dfrac{\beta}{\alpha^2}$.

   b) Determine a quadratic equation with integer coefficients which has roots

   $$\frac{\alpha}{\beta^2} \text{ and } \frac{\beta}{\alpha^2}$$

   c) Find the complex roots of the equation $x^2 + 4x + 13 = 0$.   *(AQA, 2002)*

10 Show that the equation

   $$iz + 2 = z^* + 2i$$

   has an infinite number of solutions.

8 Calculus

This chapter will show you how to

- ✦ Differentiate a polynomial function from first principles
- ✦ Recognise an improper integral
- ✦ Evaluate an improper integral where possible

Before you start

You should know how to ...	Check in
1 Find the gradient of a line joining two given points.	**1** Find the gradient of the line AB when: a) A is $(1, 3)$ and B is $(3, 7)$ b) A is $(-1, 3)$ and B is $(-2, 2)$ c) A is $(0, 5)$ and B is $(5, 5)$
2 Recognise when a fractional expression is undefined.	**2** For which values of x, if any, are the following expressions undefined? a) $\dfrac{1}{x^2}$ b) $\dfrac{1}{\sqrt{x}}$ c) $\dfrac{1}{(x-2)^2}$ d) $\dfrac{1}{x^2+4}$

3 Express x^n, where n is negative, as a fraction and vice versa.	**3** a) Express as fractions:
	    i) x^{-1} ii) x^{-2} iii) $x^{-\frac{1}{2}}$ iv) $x^{-\frac{3}{2}}$
	    b) Express as powers of x:
	    i) $\dfrac{1}{\sqrt{x}}$ ii) $\dfrac{1}{x^2\sqrt{x}}$ iii) $\dfrac{x}{\sqrt{x}}$

4 Integrate powers of x other than x^{-1}.	**4** Integrate: a) x^{-2} b) $x^{-\frac{1}{2}}$ c) $x^{-\frac{3}{2}}$ d) $\dfrac{1}{x^2\sqrt{x}}$

Links to Core modules

No knowledge of differentiation from modules C1 and C2 is assumed in section 8.1.

Knowledge of binomial expansions from module C2 would be helpful but is not assumed.

Section 8.2 assumes a knowledge of the integration contained in modules C1 and C2.

8.1 Differentiating from first principles

Quadratic functions

Consider the graph of $y = x^2$.

The gradient of the curve at the point $A(2, 4)$ is given by the gradient of the **tangent** at A.

You can find the exact value of this gradient by **differentiation from first principles.**

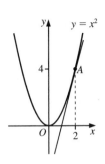

The tangent is the line that touches the graph at the point A $(2, 4)$. A close approximation to this tangent is the chord AP, where P is a point on $y = x^2$ that is very close to the point A. The x-coordinate of P is $2 + h$, where h is a very small number. As h becomes smaller and smaller, the chord AP will get closer and closer to the tangent at the point $(2, 4)$.

> The gradient of a curve at a point is discussed in module C1.

FP1

The x-coordinate of P is $2 + h$, so the y-coordinate is

$$(2 + h)^2 = 4 + 4h + h^2$$

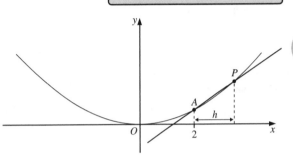

The point P is $(2 + h, 4 + 4h + h^2)$.

The chord AP will never be exactly the same as the tangent, but the slope of AP gets closer and closer to the slope of the tangent as h gets smaller and smaller.

To find the gradient of AP, you can use the formula:

$$\text{Gradient of a line joining } (x_1, y_1) \text{ to } (x_2, y_2) = \frac{y_2 - y_1}{x_2 - x_1}$$

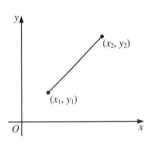

Hence:

$$\text{Gradient of } AP = \frac{y\text{-coordinate of } P - y\text{-coordinate of } A}{x\text{-coordinate of } P - x\text{-coordinate of } A}$$

$$= \frac{(2 + h)^2 - (2)^2}{(2 + h) - (2)}$$

> Substitute the coordinates of P and A.

$$= \frac{4 + 4h + h^2 - 4}{h}$$

$$= \frac{4h + h^2}{h}$$

$$= \frac{h(4 + h)}{h}$$

> Notice that you can only cancel h because h is not equal to zero.
> You cannot divide by zero.
> Instead, h is **approaching** zero.

$$= 4 + h$$

As h approaches zero (gets smaller and smaller), the gradient of AP gets closer and closer to the value 4.

When h is zero, it does not make sense to find the gradient of the line AP since A and P are the same point.

To express this mathematically, you write:

As $h \to 0$,

the gradient of chord $AP \to$ the gradient of the tangent at A

and $4 + h \to 4$.

Thus, the gradient of the tangent at A is 4.

The gradient of the curve $y = x^2$ at the point (2, 4) is 4.

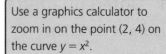

Use a graphics calculator to zoom in on the point (2, 4) on the curve $y = x^2$.

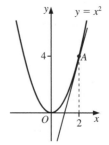

The gradient at A is 4.

FP1

Example 1

Find the gradient of the line AP, where A is the point with x-value 3 on the curve $y = x^2 + 5x + 7$, and P is the point with x-value $(3 + h)$.

Use your answer to find the equation of the tangent to the curve at the point where $x = 3$.

$\text{Gradient of } AP = \dfrac{y_2 - y_1}{x_2 - x_1}$

$$= \frac{[(3 + h)^2 + 5(3 + h) + 7] - [(3)^2 + 5(3) + 7]}{(3 + h) - 3}$$

Substituting the coordinates of P and A.

$$= \frac{9 + 6h + h^2 + 15 + 5h + 7 - (9 + 15 + 7)}{h}$$

$$= \frac{11h + h^2}{h}$$

Cancel by h ($h \neq 0$).

$$= 11 + h$$

As $h \to 0$, the gradient of AP approaches the gradient of the tangent at A. Therefore the gradient of the tangent is 11.

The tangent passes through point A and the y-coordinate of A is $9 + 15 + 7 = 31$

Therefore, the tangent is the line through (3, 31) which has gradient 11.

The equation of this tangent is:

$$y - 31 = 11(x - 3)$$
$$y = 11x - 2$$

The equation of a straight line passing through the point (x_1, y_1) is $y - y_1 = m(x - x_1)$

Example 2

Use first principles to find the gradient of the tangent to the curve $y = 2x^2 + 3x - 18$ at the point $B(a, 2a^2 + 3a - 18)$.

....

Start by finding the gradient of the chord joining point B to the point P on the curve, which has x-coordinate $a + h$.

Gradient of chord BP

$= \dfrac{[2(a + h)^2 + 3(a + h) - 18] - [2a^2 + 3a - 18]}{(a + h) - a}$

$= \dfrac{(2a^2 + 4ah + 2h^2 + 3a + 3h - 18) - (2a^2 + 3a - 18)}{h}$

$= \dfrac{4ah + 2h^2 + 3h}{h}$

$= 4a + 2h + 3$

Find the gradient of the tangent by letting h approach zero.

Hence, the gradient of the tangent at B is $4a + 3$.

> The gradient of the tangent depends on the x-value which you are considering. This is because quadratic graphs curve (they change gradient).

> As $h \to 0$, $2h \to 0$.

FP1

Exercise 8A
....

1 Find the gradient of the line AP, where A is the point with x-value 4 on the curve $y = x^2 - 11x + 4$, and P is the point with x-value $(4 + h)$.

Hence find the gradient of the tangent to the curve at the point $(4, -24)$.

2 Find the gradient of the line AP, where A is the point with x-value 2 on the curve $y = x^2 + 3x - 9$, and P is the point with x-value $(2 + h)$.

Hence find the gradient of the tangent to the curve at the point $(2, 1)$.

3 Find the gradient of the line AP, where A is the point with x-value -3 on the curve $y = 4x^2 - x + 3$, and P is the point with x-value $(-3 + h)$.

Use your answer to find the equation of the tangent to the curve at the point where $x = -3$.

4 Find the gradient of the line AP, where A is the point with x-value 5 on the curve $y = 2x^2 + 3x - 8$, and P is the point with x-value $(5 + h)$.

Use your answer to find the equation of the tangent to the curve at the point where $x = 5$.

5 Differentiate from first principles to find the gradient of the tangent to the curve $y = 3x^2 + 7x - 15$ at the point $B(a, 3a^2 + 7a - 15)$.

6 Differentiate from first principles to find the gradient of the tangent to the curve $y = 2 + 5x - 4x^2$ at the point $B(a, 2 + 5a - 4a^2)$.

. . . .

Cubic and quartic functions

You can find the gradients of cubic and quartic functions from first principles in the same way as for quadratics.

You will need to know the expansions of $(a + h)^3$ and $(a + h)^4$.

FP1

To expand $(a + h)^3$ and $(a + h)^4$, you either use the binomial theorem (see module C2), or you could repeatedly multiply $(a + h)$ by itself.

> A cubic function contains an x^3 term as its highest power, and a quartic function contains an x^4 term as its highest power.

Example 3

Find the gradient of the tangent to the curve $y = x^3$ at the point $A\ (a, a^3)$.
. . . .

Follow the same procedure as before and find the gradient of the chord AP, where P is the point $(a + h, (a + h)^3)$.

$$\text{Gradient of } AP = \frac{(a + h)^3 - a^3}{(a + h) - a}$$

> Subtract the y-coordinates and the x-coordinates.

$$= \frac{a^3 + 3a^2h + 3ah^2 + h^3 - a^3}{h}$$

> Cancel by h $(h \neq 0)$.

$$= 3a^2 + 3ah + h^2$$

As point A gets closer to point P, h approaches zero and the gradient of the chord AP approaches $3a^2$. Therefore, the gradient of the tangent to $y = x^3$ at the point A is $3a^2$.

> As $h \to 0$, $3ah$ and h^2 also approach zero.

Example 4

Expand $(2 + h)^4$. Use your answer to find the gradient of the curve $y = x^4$ at the point $A\ (2, 16)$.
. . . .

$$(2 + h)^4 = 2^4 + 4(2^3h) + \frac{4 \times 3}{2}(2^2h^2) + \frac{4 \times 3 \times 2}{3 \times 2}(2h^3) + h^4$$

> Use the binomial theorem.

$$= 16 + 32h + 24h^2 + 8h^3 + h^4$$

To find the gradient of $y = x^4$ at the point $A(2, 16)$, first find the gradient of the chord AP, where P is the point $(2 + h, (2 + h)^4)$.

$$\text{Gradient of chord } AP = \frac{(2+h)^4 - 16}{(2+h) - 2}$$

$$= \frac{16 + 32h + 24h^2 + 8h^3 + h^4 - 16}{h}$$

$$= 32 + 24h + 8h^2 + h^3$$

As $h \to 0$, point P approaches point A, and the gradient of the chord approaches the gradient of the tangent.

Therefore, the gradient of the tangent at A is 32.

> Use the expansion of $(2 + h)^4$.

> Cancel by h.

> The other three terms all approach zero as $h \to 0$.

FP1

Exercise 8B

1   Find the gradient of the line AP, where A is the point with x-value 3 on the curve $y = 2x^3 - 5x + 1$, and P is the point with x-value $(3 + h)$.

Hence, find the gradient of the tangent to the curve at the point $(3, 40)$.

2   Find the gradient of the line AP, where A is the point with x-value 1 on the curve $y = x^4 + 7$, and P is the point with x-value $(1 + h)$.

Hence, find the gradient of the tangent to the curve at the point $(1, 8)$.

3   Differentiate from first principles to find the gradient of the tangent to the curve $y = x^4 - 3x^2 + 11$ at the point $B(a, a^4 - 3a^2 + 11)$.

4   Find the gradient of the line AP, where A is the point with x-value 2 on the curve $y = x^5$, and P is the point with x-value $(2 + h)$.

Use your answer to find the equation of the tangent to the curve at the point where $x = 2$.

5   Find the gradient of the line AP, where A is the point with x-value 4 on the curve $y = \frac{1}{x}$, and P is the point with x-value $(4 + h)$.

Use your answer to find the equation of the tangent to the curve at the point where $x = 4$.

8.2   Improper integrals

There are two types of **improper** integral.

✦ The integral has ∞ or $-\infty$ as one of its limits. For example:

$$\int_{-\infty}^{\infty} \frac{1}{x^2}\,dx \text{ and } \int_{-\infty}^{-1} \frac{1}{x}\,dx \text{ are improper integrals.}$$

✦ The integrand is undefined either at one of its limits of
integration, or somewhere in between these limits. The function
may have a denominator which becomes zero for some value of x.

> The **integrand** is the expression to be integrated. For example, it is $\dfrac{1}{x^2}$ in Example 5.

$$\int_{0}^{1} \frac{1}{\sqrt{x}}\,dx \text{ is an improper integral since } \frac{1}{x} \text{ is undefined}$$

when $x = 0$.

$$\int_{-2}^{3} \frac{2}{x-1}\,dx \text{ is an improper integral since } \frac{2}{x-1} \text{ is undefined}$$

when $x = 1$.

FP1

Limits of integration include ∞ or $-\infty$

Replace ∞ (or $-\infty$) by n, and see what happens as n approaches ∞
(or $-\infty$). If the integral approaches a finite answer, then the integral
can be found (see Example 5). Otherwise, the integral **cannot be
found** (see Example 6).

Example 5

Determine $\displaystyle\int_{1}^{\infty} \frac{1}{x^2}\,dx$.

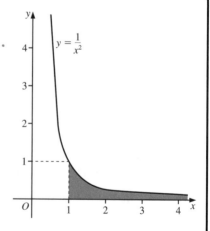

Substitute n for ∞:

$$\int_{1}^{n} \frac{1}{x^2}\,dx = \int_{1}^{n} x^{-2}\,dx$$

$$= \left[\frac{x^{-1}}{-1} \right]_{1}^{n}$$

$$= \left[\frac{-1}{x} \right]_{1}^{n}$$

$$= -\frac{1}{n} + 1$$

As $n \to \infty$, the value of the integral approaches 1. Therefore, the
improper integral $\displaystyle\int_{1}^{\infty} \frac{1}{x^2}\,dx$ can be found, and the answer is 1.

> The shaded region has a finite area even though the boundary is of infinite length.

Example 6

Determine whether the integral $\int_1^\infty \dfrac{1}{\sqrt{x}}\,dx$ has a value. If so, find the value of the integral.

· ·

Again, substitute n for ∞:

$$\int_1^n \frac{1}{\sqrt{x}}\,dx = \int_1^n x^{-\frac{1}{2}}\,dx$$

$$= \left[2x^{\frac{1}{2}}\right]_1^n$$

$$= 2\sqrt{n} - 2$$

As $n \to \infty$, the value of this integral does not approach a finite number, and so the integral cannot be found.

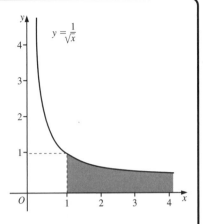

The shaded region does not have a finite area, although it looks very similar to the one in Example 5.

FP1

Integral undefined at a particular value

If the integrand is undefined at one of the limits of integration, replace that limit of integration with p, and see what happens as p approaches the limit of integration. If the integral approaches a finite answer, then the integral can be found.

Example 7

Determine $\int_0^1 \dfrac{1}{\sqrt{x}}\,dx$.

· ·

This is an improper integral since the integral is undefined when $x = 0$.

Replace the lower limit of integration by p.

$$\int_p^1 \frac{1}{\sqrt{x}}\,dx = \int_p^1 x^{-\frac{1}{2}}\,dx$$

$$= \left[2x^{\frac{1}{2}}\right]_p^1$$

$$= 2 - 2\sqrt{p}$$

As p approaches zero, the value of $2 - 2\sqrt{p}$ approaches 2.

Therefore the improper integral $\int_0^1 \dfrac{1}{\sqrt{x}}\,dx$ can be found, and its value is 2.

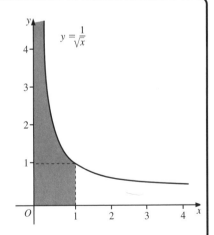

If the integral does not approach a finite answer, the integral cannot be found.

Example 8

Determine $\int_{-1}^{0} \dfrac{1}{x^2}\, dx$.

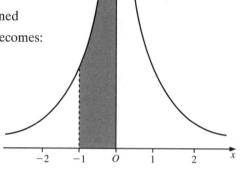

This is an improper integral since the integrand $\dfrac{1}{x^2}$ is undefined when $x = 0$. Replace the upper limit by p, and the integral becomes:

$$\int_{-1}^{p} \dfrac{1}{x^2}\, dx = \int_{-1}^{p} x^{-2}\, dx$$

$$= \left[\dfrac{x^{-1}}{-1}\right]_{-1}^{p}$$

$$= \left[\dfrac{-1}{x}\right]_{-1}^{p}$$

$$= \dfrac{-1}{p} - \dfrac{-1}{-1}$$

$$= -\dfrac{1}{p} - 1$$

As p approaches zero, $-\dfrac{1}{p} - 1$ does not approach a finite number.

Therefore the improper integral $\int_{-1}^{0} \dfrac{1}{x^2}\, dx$ cannot be found.

If the integrand is undefined at some point between the two limits of integration, then you need to split the integral into two parts, one to the left, and one to the right of the point where the integrand is undefined and use a similar approach.

Example 9

Determine $\int_{-4}^{2} \dfrac{1}{x^2}\, dx$.

Write $\int_{-4}^{2} \dfrac{1}{x^2}\, dx = \int_{-4}^{0} \dfrac{1}{x^2}\, dx + \int_{0}^{2} \dfrac{1}{x^2}\, dx$

Now calculate the two integrals on the right using the method of Example 8. You will discover that neither of the two integrals can be found, and therefore the integral $\int_{-4}^{2} \dfrac{1}{x^2}\, dx$ cannot be found.

Notice that the integrand is undefined when $x = 0$.

Split the integral into two parts from -4 to 0 and 0 to 2.

FP1

Exercise 8C

. . . .

For each of these improper integrals, decide whether the integral can be found. If it can, give the answer.

1 $\int_0^1 \dfrac{1}{\sqrt[3]{x}}\,dx$ **2** $\int_0^2 \dfrac{1}{x^{\frac{3}{2}}}\,dx$ **3** $\int_0^\infty \dfrac{1}{\sqrt{x}}\,dx$

4 $\int_0^\infty \dfrac{x}{x^3}\,dx$ **5** $\int_0^2 \dfrac{x^{\frac{1}{3}}}{x^{\frac{5}{2}}}\,dx$ **6** $\int_0^1 \dfrac{x-1}{x^5}\,dx$

. . . .

Summary

FP1

You should know how to …	Check out
1 Differentiate a polynomial function from first principles.	**1** Find, from first principles, the gradient of the curve $y = 2x^3 + x$ at the point where $x = 3$.
2 Recognise an improper integral.	**2** Which of these are improper integrals? a) $\int_0^1 \dfrac{1}{x^{\frac{1}{2}}}\,dx$ b) $\int_1^4 \dfrac{1}{x^{\frac{1}{2}}}\,dx$ c) $\int_1^\infty \dfrac{1}{x^{\frac{1}{2}}}\,dx$
3 Evaluate an improper integral where possible.	**3** Evaluate, where possible: a) $\int_0^1 \dfrac{1}{x^{\frac{1}{2}}}\,dx$ b) $\int_0^1 \dfrac{1}{x^2}\,dx$ c) $\int_9^\infty \dfrac{1}{x^{\frac{1}{2}}}\,dx$ d) $\int_9^\infty \dfrac{1}{x^{\frac{3}{2}}}\,dx$

Revision exercise 8

1 Given that $f(x) = 2x^2 + 1$:
   a) show that $f(2 + h) - f(2) = 8h + 2h^2$
   b) hence find the value of $f'(2)$.

2 Given that $f(x) = x^3 - x$:
   a) show that $f(3 + h) - f(3) = 26h + 9h^2 + h^3$
   b) hence find the value of $f'(3)$.

3 Given that $f(x) = 2x^4 + 1$:
   a) write down the value of $f(-1)$
   b) show that $f(-1 + h) = 3 - 8h + 12h^2 - 8h^3 + 2h^4$
   c) hence, find the value of $f'(-1)$.

4 It is given that $f(x) = x^2 + x$ and $g(x) = 2(x^2 + x)$.

   a) Express $f(1 + h) - f(1)$ in terms of h.

   b) Hence, show that $g(1 + h) - g(1) = 6h + 2h^2$.

   c) Use your results in parts a) and b) to find the value of $f'(1)$ and $g'(1)$.

5 It is given that $f(x) = x^2 - x$ and $g(x) = f(2x)$.

   a) Express $f(2 + h) - f(2)$ in terms of h.

   b) Hence show that $g(1 + h) - g(1) = 6h + 4h^2$.

   c) Use your results in parts a) and b) to find the value of $f'(2)$ and $g'(1)$.

6 Evaluate, where possible:

   a) $\displaystyle\int_0^1 x^{\frac{1}{3}}\, dx$ 　　　　　　 b) $\displaystyle\int_1^\infty x^{\frac{1}{3}}\, dx$

7 Evaluate, where possible:

   a) $\displaystyle\int_0^1 x^{-\frac{1}{3}}\, dx$ 　　　　　　 b) $\displaystyle\int_1^\infty x^{-\frac{1}{3}}\, dx$

8 Evaluate, where possible:

   a) $\displaystyle\int_0^1 x^{-\frac{4}{3}}\, dx$ 　　　　　　 b) $\displaystyle\int_1^\infty x^{-\frac{4}{3}}\, dx$

9 Evaluate, where possible:

   a) $\displaystyle\int_0^1 \left(5x^{\frac{2}{3}} - 3x^{-\frac{2}{3}}\right) dx$ 　　 b) $\displaystyle\int_{81}^\infty \left(x^{-\frac{5}{4}} + 3x^{-\frac{7}{4}}\right) dx$

10 a) Sketch the graph of $y = \dfrac{1}{x^2}$.

   b) Indicate on your diagram the region between the graph and the x-axis to the left of $x = -1$.

   c) Show by integration that the area of this region is 1.

   d) Explain briefly what is meant by saying that the integral used in part c) is an improper integral.

9 Trigonometry

This chapter will show you how to

✦ Find the general solutions of trigonometric equations

Before you start

You should know how to ...	Check in
1 Use degrees and radians for angles.	**1** Write the radian equivalents of: a) 45° b) 360° c) 120° d) 30°
2 Use exact values for the sine, cosine and tangent of 30°, 45° and 60°.	**2** Give the exact values of: a) $\sin \dfrac{\pi}{3}$ b) $\cos \dfrac{\pi}{4}$ c) $\sin \dfrac{2\pi}{3}$

FP1

Links to Core modules

It is assumed that you are familiar with the use of radians from module C2. The solving of trigonometric equations is explained fully in this chapter, though it would be helpful if you have studied the techniques included in module C2.

You can find the sine, cosine and tangent of **any** angle, not just an angle in the range 0–90°.

This enables you to find many solutions to trigonometric equations.

The usual, and the simplest, method of finding more than one solution of a trigonometric equation is to use the **general solution**.

9.1 General solutions of equations involving cosines

To find the general solution for the equation $\cos \theta = \frac{1}{2}$, look at the graphs of $y = \cos \theta$ and $y = \frac{1}{2}$.

> See Module C2 for the graph of $y = \cos \theta$.

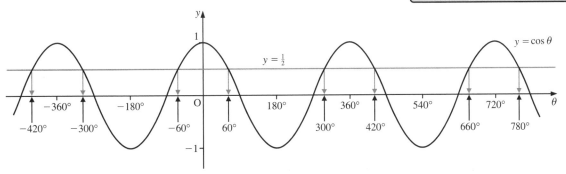

You can see that the solutions for θ are:

$$..., -420°, -300°, -60°, 60°, 300°, 420°, 660°, 780°, ...$$

You can write all these solutions for θ as $\theta = 360n° \pm 60°$, where n is an integer.

When $n = 2$, for example, $360n \pm 60°$ takes the values $720° \pm 60°$, that is $780°$ and $660°$, and so on.

If radians are used instead of degrees, the general solution of the equation $\cos \theta = \frac{1}{2}$ is $\theta = 2\pi n \pm \frac{\pi}{3}$.

This result can be generalised.

> The general solution of the equation $\cos \theta = \cos \alpha$ for any angle α is:
>
> $$\theta = 360n° \pm \alpha$$
>
> if α is given in degrees.
>
> If θ and α are measured in radians, this general solution becomes:
>
> $$\theta = 2n\pi \pm \alpha$$

When you evaluate $\cos^{-1} \frac{1}{2}$ on your calculator, you should get $60°$. Your calculator will give you a single (usually acute) angle but there are infinitely many solutions.

Note that the integer n can be negative.

Since n can have an infinite number of values, trigonometric equations have an infinite number of solutions.

Example 1

Find the general solution, in degrees, of the equation $\cos \theta = -\frac{1}{2}$.

....

$-\frac{1}{2}$ is the cosine of $120°$.

So, the general solution is:

$$\theta = 360n° \pm 120°$$

In the general solution, the angle α can be obtuse.

Example 2

Find the general solution, in degrees, of the equation $\cos 5\theta = \frac{\sqrt{3}}{2}$.

....

$\dfrac{\sqrt{3}}{2}$ is the cosine of $30°$.

So, the general solution for 5θ is:

$$5\theta = 360n° \pm 30°$$

Dividing by 5, the general solution for θ is:

$$\theta = 72n° \pm 6°$$

Remember to find the **general solution for 5θ first**. Then transform the equation to give the general solution for θ.

You can check these values on a graphics calculator, after selecting the correct **range** or **view window**.

FP1

Example 3

Find the general solution, in radians, of the equation $\cos 3x = \dfrac{1}{\sqrt{2}}$.

$\dfrac{1}{\sqrt{2}}$ is the cosine of $\dfrac{\pi}{4}$ radians.

So the general solution for $3x$ is $\quad 3x = 2n\pi \pm \dfrac{\pi}{4}$.

and the general solution for x is:

$$x = \frac{2}{3}n\pi \pm \frac{\pi}{12}$$

Example 4

Find the general solution, in radians, of the equation
$\cos\left(2x - \dfrac{\pi}{6}\right) = \dfrac{1}{2}$.

$\dfrac{1}{2}$ is the cosine of $\dfrac{\pi}{3}$ radians.

The general solution for $\left(2x - \dfrac{\pi}{6}\right)$ is:

$$2x - \frac{\pi}{6} = 2n\pi \pm \frac{\pi}{3}$$

$$2x = 2n\pi \pm \frac{\pi}{3} + \frac{\pi}{6}$$

The general solution for x is:

$$x = n\pi \pm \frac{\pi}{6} + \frac{\pi}{12}$$

Example 5

Find the general solution, in degrees, of the equation $2\cos^2 x = \cos x$.

First, rearrange the equation:

$$2\cos^2 x - \cos x = 0$$

$$\cos x\,(2\cos x - 1) = 0$$

$$\cos x = 0 \quad \text{or} \quad \cos x = \tfrac{1}{2}$$

$$0 = \cos 90° \quad \text{and} \quad \tfrac{1}{2} = \cos 60°.$$

> One of the factors must be zero.

Hence, the general solution is:

$$x = 360n° \pm 90° \text{ and } 360n° \pm 60°$$

Exercise 9A

In Questions **1** to **6** find the general solution of each equation:

a) in degrees   b) in radians

1 $\cos x = \dfrac{\sqrt{3}}{2}$

2 $\cos x = \dfrac{1}{2}$

3 $\cos 2\theta = \dfrac{1}{\sqrt{2}}$

4 $\cos 3\theta = -\dfrac{\sqrt{2}}{2}$

5 $\cos 5x = 1$

6 $4\cos^2 x = 1$

In questions **7–10** find the general solution of each equation in radians.

7 $\cos\left(3\theta - \dfrac{\pi}{3}\right) = \dfrac{1}{2}$

8 $\cos\left(2\theta - \dfrac{\pi}{4}\right) = \dfrac{\sqrt{3}}{2}$

9 $\cos\left(4\theta - \dfrac{\pi}{6}\right) = \dfrac{1}{\sqrt{2}}$

10 $\cos(2\theta - 1) = -0.2$

FP1

9.2 General solutions of equations involving sines

In the same way as for cosine curves, you can find the general solution of $\sin\theta = \frac{1}{2}$ by looking at the graphs of $y = \sin\theta$ and $y = \frac{1}{2}$.

> See Module C2.

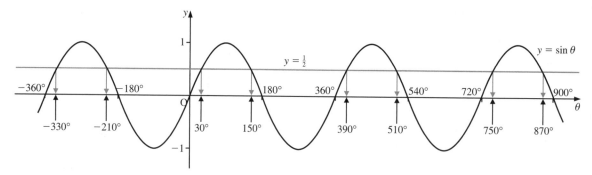

You can see that the solutions for θ are:

  ..., $-330°$, $-210°$, $30°$, $150°$, $390°$, $510°$, $750°$, ...

You can write these solutions as:

  ..., $-360° + 30°$, $-180° - 30°$, $30°$, $180° - 30°$, $360° + 30°$,
  $540° - 30°$, $720° + 30°$, ...

So, the values for θ can be written as $180n° + (-1)^n\,30°$ for any integer n, as the table shows.

> As in the cosine solution, n can take negative values. For example, when $n = -1$, $180n° + (-1)^n 30°$ takes the value $-180° - 30° = -210°$.

n	$180n° + (-1)^n 30°$
0	$0 + 30° = 30°$
1	$180° - 30° = 150°$
2	$360° + 30° = 390°$
3	$540° - 30° = 510°$

If radians are used instead of degrees, the general solution of the equation $\sin \theta = \frac{1}{2}$ is:

$$\theta = \pi n + (-1)^n \frac{\pi}{6}$$

This result can be generalised.

> The general solution of the equation $\sin \theta = \sin \alpha$ for any angle α is:
> $$\theta = 180n° + (-1)^n \alpha$$
> if α is given in degrees.
> If θ and α are measured in radians, this general solution becomes
> $$\theta = n\pi + (-1)^n \alpha$$

FP1

Example 6

Find the general solution, in degrees, for the equation $\sin \theta = \dfrac{\sqrt{3}}{2}$.

$\dfrac{\sqrt{3}}{2} = \sin 60°$.

So the general solution is:

$$\theta = 180n° + (-1)^n 60°$$

> Standard angle facts are given on page 42.

Example 7

Find the general solution, in degrees, for the equation $\sin 3\theta = \dfrac{1}{\sqrt{2}}$.

$\dfrac{1}{\sqrt{2}} = \sin 45°$.

So the general solution for 3θ is:
$$3\theta = 180n° + (-1)^n 45°$$
and the general solution for θ is:
$$\theta = 60n° + (-1)^n 15°$$

> **Remember** to find the general solution for 3θ first.

Example 8

Find the general solution, in radians, for the equation $\sin 4\theta = 0.7$.

Using radians, the calculator gives:

$$\sin^{-1} 0.7 = 0.775\,397\ldots$$
$$\therefore 4\theta = n\pi + (-1)^n\, 0.775\,397\ldots$$
$$\theta = n\frac{\pi}{4} + (-1)^n\, 0.193\,849\ldots$$
$$\approx n\frac{\pi}{4} + (-1)^n\, (0.194) \quad \text{(To 3 significant figures)}$$

Example 9

Find the general solution, in degrees, of the equation $4 \sin^2 \theta = 1$.

Rearrange the equation $4 \sin^2 \theta = 1$ as:

$$4 \sin^2 \theta - 1 = 0$$
$$(2 \sin \theta - 1)(2 \sin \theta + 1) = 0$$
$$\sin \theta = \tfrac{1}{2} \quad \text{or} \quad -\tfrac{1}{2}$$
$$= \sin 30° \quad \text{or} \quad \sin (-30°)$$

> Factorise, using the difference of two squares.

So, the general solution is $\theta = 180n° + (-1)^n 30°$ and
$\theta = 180n° + (-1)^{n+1} 30°$.

> The power of -1 goes up by 1 because of the $-30°$.

You can combine these to give $\theta = 180n° \pm 30°$.

FP1

Example 10

Find the general solution, in radians, of the equation
$2 \cos^2 x = 3 \sin x$.

Transform the equation, using the identity $\cos^2 x + \sin^2 x \equiv 1$.

> The identity $\cos^2 x + \sin^2 x \equiv 1$ is in module C2.

So, $2 \cos^2 x = 3 \sin x$ becomes:

$$2(1 - \sin^2 x) = 3 \sin x$$
$$2 \sin^2 x + 3 \sin x - 2 = 0$$
$$(2 \sin x - 1)(\sin x + 2) = 0$$

So, $\sin x = -2$ or $\sin x = \tfrac{1}{2}$.

However, $\sin x$ is not equal to -2 for any value of x.

> The minimum value of $\sin x = -1$.

The only possibility is $\sin x = \tfrac{1}{2}$, which gives:

$$x = n\pi + (-1)^n \frac{\pi}{6}$$

> $\sin^{-1} \dfrac{1}{2} = \dfrac{\pi}{6}$

Exercise 9B

In Questions **1** to **6** find the general solution of each equation:

a) in degrees   b) in radians

1 $\sin x = \dfrac{\sqrt{3}}{2}$

2 $\sin x = \dfrac{1}{\sqrt{2}}$

3 $\sin 2\theta = -\dfrac{1}{2}$

4 $\sin 3\theta = \dfrac{\sqrt{2}}{2}$

5 $\sin 5x = -1$

6 $\sin^2 x = \dfrac{3}{4}$

In Questions **7** to **13**, find the general solution of each equation in radians.

7 $\sin\left(3\theta - \dfrac{\pi}{3}\right) = \dfrac{1}{2}$

8 $\sin\left(2\theta - \dfrac{\pi}{4}\right) = \dfrac{\sqrt{3}}{2}$

9 $\sin(2\theta - 1) = -0.4$

10 $2\cos^2 x = 3\sin x + 3$

11 $4\cos^2 x = 5 - 5\sin x$

12 $6\cos^2 x = 5 + \sin x$

13 $\sin^2 3x + \cos 3x + 1 = 0$

9.3 General solutions of equations involving tangents

To find the general solution of $\tan\theta = 1$, look at the graphs of $y = \tan\theta$ and $y = 1$.

FP1

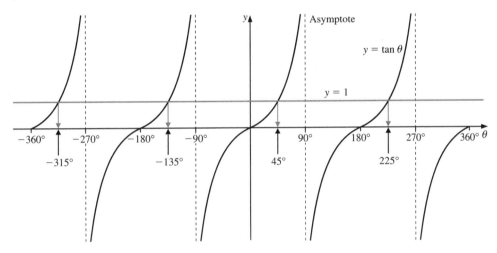

You can see that the solutions for θ are:

..., $-315°$, $-135°$, $45°$, $225°$, ...

or $180n° + 45°$ for any integer, n.

If radians are used instead of degrees, the general solution of the equation $\tan\theta = 1$ is:

$$\theta = \pi n + \frac{\pi}{4}$$

This can be generalised.

> The general solution of the equation $\tan\theta = \tan\alpha$ for any angle α is:
>
> $$\theta = 180n° + \alpha$$
>
> if α is given in degrees.
>
> If θ and α are measured in radians, this general solution is:
>
> $$\theta = n\pi + \alpha$$

Example 11

Find the general solution, in degrees, of the equation $\tan 4\theta = -\sqrt{3}$.

You know that $\sqrt{3}$ is $\tan 60°$. So, $\tan(-60°) = -\sqrt{3}$.

The general solution is:

$$4\theta = 180n° - 60°$$
$$\therefore \quad \theta = 45n° - 15°$$

Example 12

Find the general solution, in radians, of the equation

$$\tan\left(\frac{\pi}{4} - 3x\right) = \sqrt{3}.$$

$$\sqrt{3} = \tan\frac{\pi}{3}$$

In this problem, you need the general solution for $\frac{\pi}{4} - 3x$.

$$\frac{\pi}{4} - 3x = n\pi + \frac{\pi}{3}$$

$$3x - \frac{\pi}{4} = -n\pi - \frac{\pi}{3}$$

$$3x = -n\pi - \frac{\pi}{3} + \frac{\pi}{4}$$

$$3x = -n\pi - \frac{\pi}{12}$$

$$x = -\frac{n}{3}\pi - \frac{\pi}{36}$$

So, you can write the general solution as:

$$x = \frac{m\pi}{3} - \frac{\pi}{36}$$

> Multiply the equation by -1.

> For any integer n, $-\dfrac{n}{3}$ can be written as $\dfrac{m}{3}$, where $m = -n$ and is also any integer. This is a neater way to write the solution.

Example 13

Find the general solution, in radians, of the equation $\tan(2x - 1) = -0.2$.

Using radians, the calculator gives $\tan^{-1}(-0.2) = -0.197\,39\ldots$

$$2x - 1 = n\pi - 0.197\,39\ldots$$
$$2x = n\pi + 0.802\,61\ldots$$
$$x = \tfrac{1}{2}n\pi + 0.401 \quad \text{(to 3 significant figures)}$$

FP1

Example 14

Find the general solution, in radians, of the equation
$$\sin\left(2x + \frac{\pi}{4}\right) = \cos\left(2x + \frac{\pi}{4}\right).$$

....

Divide both sides of the equation by $\cos\left(2x + \frac{\pi}{4}\right)$.

This gives $\tan\left(2x + \frac{\pi}{4}\right) = 1$. Therefore:

$$2x + \frac{\pi}{4} = n\pi + \frac{\pi}{4}$$

$$2x = n\pi$$

$$x = \frac{n\pi}{2}$$

> You need to simplify the equation so that it contains only one type of trigonometrical function.

> The identity $\tan x \equiv \dfrac{\sin x}{\cos x}$ is explained in module C2.

> $\tan\dfrac{\pi}{4} = 1.$

FP1

Exercise 9C

In Questions **1** to **6**, find the general solution of each equation:

a) in degrees   b) in radians.

1 $\tan x = \sqrt{3}$

2 $\tan x = -1$

3 $\tan 6\theta = 1$

4 $\tan 3\theta = -\dfrac{\sqrt{3}}{3}$

5 $\tan 5x = 0.6$

6 $3\tan^2 x = 0.8$

In Questions **7** to **13**, find the general solution of each equation in radians.

7 $\tan\left(3\theta - \dfrac{\pi}{3}\right) = 1$

8 $\tan\left(2\theta - \dfrac{\pi}{4}\right) = \sqrt{3}$

9 $\tan\left(4\theta + \dfrac{\pi}{6}\right) = -\dfrac{1}{\sqrt{3}}$

10 $\sin 4x = \cos 4x$

11 $\sin\left(x + \dfrac{\pi}{6}\right) = \cos\left(x + \dfrac{\pi}{6}\right)$

12 $\sin(3x - 0.2) = \cos(3x - 0.2)$

13 $\sin(4x + 0.7) = \cos(4x + 0.7)$

Summary

You should know how to ...	Check out
1 Find general solutions for equations involving cosines.	**1** Find the general solution, in degrees, of the equation $\cos(2x - 20°) = -0.5$.
2 Find general solutions for equations involving sines.	**2** Find the general solution, in radians, of the equation $\sin\left(3x + \dfrac{\pi}{3}\right) = \dfrac{\sqrt{3}}{2}$.
3 Find general solutions for equations involving tangents.	**3** Find the general solution, in degrees, of the equation $\tan(10x - 5°) = 0.9$.

FP1 : Revision exercise 9

1 Write down the general solutions of the equations $\sin x = 0$, $\cos x = 0$ and $\tan x = 0$. Give your answers a) in degrees, and b) in radians.

2 Find the general solutions of the equations $\sin x = \frac{1}{2}$, $\cos x = \frac{1}{2}$, $\tan x = 1$. Give your answers a) in degrees, and b) in radians.

3 Find the general solutions of the equations $2\sin 2x = \sqrt{3}$, $2\cos 2x = \sqrt{3}$, $\tan 2x = \sqrt{3}$. Give your answers a) in degrees, and b) in radians.

4 Find the general solutions of the equations $\sin\left(x + \dfrac{\pi}{4}\right) = -\dfrac{1}{2}$,

$\cos\left(x - \dfrac{\pi}{4}\right) = -\dfrac{1}{2}$ and $\tan\left(x - \dfrac{\pi}{4}\right) = -1$.

Give your answers in radians.

5 Find the general solution of the equation $\tan x = -\sqrt{3}$, leaving your answers in terms of π.   *(AQA, 2004)*

6 Given that

$$\sqrt{3}\sin\theta + \cos\theta = 0$$

a) find the value of $\tan\theta$

b) find the general solution for θ.   *(AQA/NEAB, 1998)*

7 Find the general solution of the equation

$$\cos(3x - 20°) = -0.2,$$

giving numerical values of angles to the nearest 0.1°.   *(AQA, 2003)*

8 Find the general solution of the equation

$$5 \tan (3x + 30°) = 2$$

giving numerical values of angles to the nearest degree.   *(AQA, 2002)*

9 Find the general solution of the equation

$$\cos\left(x + \frac{\pi}{6}\right) = -0.5$$

giving your answers in terms of π.   *(AQA, 2002)*

10 a) Sketch the graph of $y = \tan x$.

   b) Find the general solution of the equation $\tan x = \sqrt{3}$.   *(AQA, 2003)*

FP1

10 Numerical solution of equations

This chapter will show you how to

- ✦ Locate a root of an equation in an interval by considering a change of sign
- ✦ Find numerical approximations for roots of equations, using three different techniques
- ✦ Use a step-by-step method to solve differential equations

Before you start

You should know how to ...	Check in
1 Recognise similar triangles and use the ratio of their side lengths.	**1** In this diagram the lines AB and CD are parallel. Calculate the length of BD. NOT TO SCALE
2 Differentiate polynomial functions.	**2** Differentiate: a) x^3 b) $5x^2 + 2x$ c) $7x^4 - 10$

Links to Core modules

The differentiation needed in this chapter is in module C1.

Locating a root is in the C3 module but is fully explained in this chapter. For examination purposes, you do not need to know how to justify the techniques introduced here, but it may be helpful to have a grasp of why they work.

Logarithmic functions appear to a small extent in section 10.3, and are explained in module C2.

However, natural logarithms are not assessed in the FP1 module.

10.1 Location of a root within an interval

Many equations cannot be solved using algebraic procedures to give exact solutions, and so you use **numerical methods** to solve them.

First rearrange the equation, if necessary, into the form $f(x) = 0$.

For example, the equation $x^3 + 5x = 9$ can be expressed as $f(x) = 0$ where:

$$f(x) = x^3 + 5x - 9$$

$f(x) = x^3 + 5x - 9$ is an example of a **function**.

Then see if you can find two numbers a and b such that $f(a)$ is negative and $f(b)$ is positive, or vice versa. In the example, $f(1) = -3$ and $f(2) = 9$.

Since the function f is continuous for all values of x, you can be certain that $f(x)$ will be 0 for some value of x in the interval $1 < x < 2$. Therefore, you have located a root within an interval.

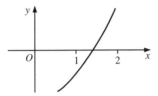

If necessary you can locate the root within smaller intervals by trying, for example, $f(1.1)$, $f(1.2)$, ...

If f were not continuous, you could have a situation similar to that shown in the graph on the right, where $f(-1)$ and $f(1)$ are of opposite signs and yet $f(x)$ is not zero for any value between -1 and 1.

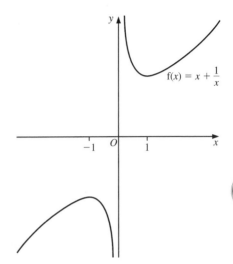

FP1

10.2 Interval bisection

As the name suggests, if you know that there is a root of $f(x) = 0$ between $x = a$ and $x = b$, try $x = \dfrac{(a+b)}{2}$. The sign of $f\left(\dfrac{a+b}{2}\right)$ determines on which side of $\dfrac{(a+b)}{2}$ the root lies.

Take, for example, the function $f(x)$.

$f(x)$ has a root between a and b.

$f(a)$ is positive and $f(b)$ is negative.

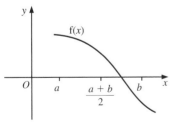

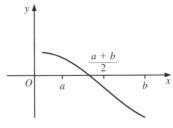

If $f\left(\dfrac{a+b}{2}\right)$ is positive, the root is between $\dfrac{a+b}{2}$ and b.

If $f\left(\dfrac{a+b}{2}\right)$ is negative, the root is between $\dfrac{a+b}{2}$.

The method is repeated until you find an answer to the degree of accuracy required.

> It is tempting to take short cuts if one of the values is very close to zero. However, repeated interval bisection is a particularly useful algorithm for a computer program.

Example 1

Show that the equation $x^3 + 5x = 9$ has a root between 1 and 2.
Use the method of interval bisection twice to obtain an interval of
width 0.25 within which the root must lie.

$$f(x) = x^3 + 5x - 9$$
$$f(1) = 1 + 5 - 9 = -3$$
$$f(2) = 8 + 10 - 9 = 9$$

$f(1)$ is negative and $f(2)$ is positive.

Therefore, a root lies between 1 and 2.

Now put $x = 1.5$, which gives:

$$f(1.5) = 1.875$$

$f(1.5)$ is positive and $f(1)$ is negative. Therefore, a root lies between
1 and 1.5.

Now consider $f(1.25)$, which is $-0.796\,875$.

Hence, $f(1.25)$ and $f(1.5)$ have opposite signs, and you can conclude
that there is a root between 1.25 and 1.5.

> It is important to mention the
> change of sign when showing
> that there is a root between two
> values of x.

FP1

> You can use the zoom and trace
> facilities on a graphics calculator
> to verify the location of the root.

Example 2

Show that the equation $x - 7\log_{10}x = 0$ has a root between 4 and 5.

Use the method of interval bisection twice to obtain an interval of
width 0.25 within which the root must lie.

$$\text{Let } f(x) = x - 7\log_{10}x$$
$$f(4) \approx -0.214, \text{ and } f(5) \approx 0.107.$$

$f(4)$ is negative and $f(5)$ is positive.

Therefore, a root lies between 4 and 5.

Now put $x = 4.5$ (the mid-value of 4 and 5),

which gives:

$$f(4.5) \approx -0.072$$

hence, a root lies between 4.5 and 5.

Let $x = 4.75$:

$$f(4.75) = +0.013\,14\ldots$$

Hence, a root lies between 4.5 and 4.75.

Exercise 10A

In each of Questions **1** to **6**, show that the given equation has a root in the given interval. Then use the method of interval bisection twice to obtain a smaller interval within which the root must lie.

1 Equation $x^3 - 4x = 5$; interval $2 < x < 3$.

2 Equation $\sin \dfrac{\pi x}{2} = 3x - 1$; interval $0 < x < 1$.

3 Equation $\sin 3x = x^2$; interval $0.6 < x < 1$.

4 Equation $x^3 - 5x^2 + 7x - 5 = 0$; interval $3 < x < 4$.

5 Equation $x^2 + 4\log_{10} x = 0$; interval $0.5 < x < 1$.

6 Equation $\sin x + \ln x = 0$; interval $0.4 < x < 0.6$.

> In questions 2, 3 and 6, ensure that your calculator is in **radian mode**.

> $\ln x$ is the natural logarithm of x. You should find this function on your calculator. It will not be examined in the FP1 module.

FP1

10.3 Linear interpolation

In solving the equation $f(x) \equiv x^3 + 5x - 9$, you can deduce from $f(1) = -3$ and $f(2) = 9$ that the root of $f(x) \equiv x^3 + 5x - 9 = 0$ is likely to be much nearer to 1 than to 2, since $|f(2)| > |f(1)|$.

> $|f(1)|$ means the **modulus** of $f(1)$.
> $|f(1)| = 3$.

This intuitive approach is formalised in **linear interpolation**, where the two points $(1, -3)$ and $(2, 9)$ are joined by a straight line and the x-value of the point where this line crosses the x-axis is calculated.

The diagram shows the estimated root at $x = 1 + p$.

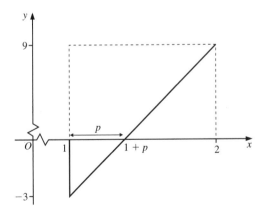

Using similar triangles:

$$\frac{p}{3} = \frac{1}{9 + 3}$$

from which you can quickly obtain the estimated root as 1.25.

Example 3

Show that the equation $\cos\dfrac{\pi x}{2} = 4x - 1$ has a root between 0 and 1.

Use linear interpolation to find an approximate value of this root.

$$\cos\frac{\pi x}{2} = 4x - 1$$

Hence:

$$\cos\frac{\pi x}{2} - 4x + 1 = 0$$

Let $f(x) = \cos\dfrac{\pi x}{2} - 4x + 1$, which gives:

$$f(0) = 2$$

$$f(1) = -3$$

The change of sign indicates that there is a root of $f(x) = 0$ between $x = 0$ and $x = 1$.

Using similar triangles, with the root at $x = p$:

$$\frac{p}{2} = \frac{1}{2+3}$$

$$p = 0.4$$

So, the required approximate value is 0.4.

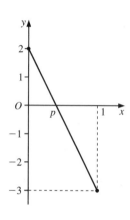

FP1

Exercise 10B

In each of Questions **1** to **6**, show that the given equation has a root in the given interval. Then use the method of linear interpolation once to obtain an approximate value for the root.

1 Equation $x^3 = 7 - 5x$; interval $1 < x < 2$.

2 Equation $\sin\dfrac{\pi x}{2} = 4x - 1$; interval $0 < x < 1$.

> In Questions 2–5, ensure that your calculator is set to radian mode.

3 Equation $\tan x + 1 - 4x^2 = 0$; interval $1.42 < x < 1.44$.

4 Equation $\sin x - 0.2x = 0$; interval $2.5 < x < 3$.

5 Equation $\sin x + \log_{10} x = 0$; interval $0.3 < x < 0.5$.

6 Equation $x - \ln x = 2$; interval $3 < x < 4$.

> In x will not be assessed in the FP1 examination.

10.4 The Newton–Raphson method

If α is an approximation to a root of $f(x) = 0$, then $\alpha - \dfrac{f(\alpha)}{f'(\alpha)}$ is generally a better approximation.

Consider the graph of $y = f(x)$. Draw the tangent at P, where $x = \alpha$, and let the tangent meet the x-axis at T.

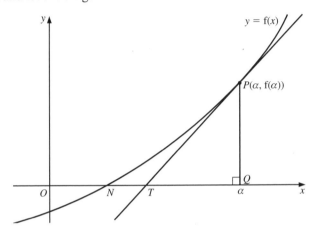

You can see that the x-value at T is closer than α is to the x-value at N, where the graph cuts the axis.

> So the x-value at T is closer to the root of $f(x) = 0$ than α is.

Using the triangle PTQ gives:

$$\text{Gradient of tangent} = \frac{PQ}{QT}$$

$$f'(\alpha) = \frac{f(\alpha)}{QT}$$

> Gradient of tangent $= f'(\alpha)$ and $PQ = f(\alpha)$.

$$QT = \frac{f(\alpha)}{f'(\alpha)}$$

The x-value of the point T is: $\alpha - QT = \alpha - \dfrac{f(\alpha)}{f'(\alpha)}$

which is a better approximation to the root of $f(x) = 0$.

When the root of $f(x) = 0$ is not close to α, the method may fail.

The next x-value found is at T, which is further from the root than α is.

In this example, $f'(\alpha) = 0$.

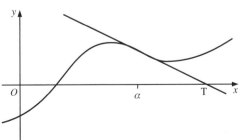

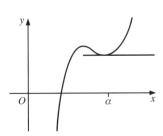

In its iterative form, the **Newton–Raphson method** for solving $f(x) = 0$ gives:

$$x_{n+1} = x_n - \frac{f(x_n)}{f'(x_n)}$$

Example 4

Use the Newton–Raphson method once, with an initial value of $x = 3$, to find an approximation for a root of the equation

$$14 + 8x + 5x^2 - x^4 = 0$$

Give your answer to three significant figures.

Write $f(x) = 14 + 8x + 5x^2 - x^4$.

Differentiation gives:

$$f'(x) = 8 + 10x - 4x^3$$

Putting $x = 3$:

$$f(3) = 2$$
$$f'(3) = -70$$

Using Newton–Raphson, the required approximation is:

$$3 - \frac{f(3)}{f'(3)} = 3 - \frac{2}{-70}$$
$$= 3.03 \text{ (to 3 significant figures)}$$

Example 5

Use the Newton–Raphson method, with an initial value of $x = 1$, to find an approximation for a root of the equation $x^3 + 5x - 9 = 0$.

Write $f(x) = x^3 + 5x - 9$.

Differentiation gives:

$$f'(x) = 3x^2 + 5$$

Putting $x = 1$:

$$f(1) = -3$$
$$f'(1) = 8$$

Using Newton–Raphson, the required approximation is:

$$1 - \frac{f(1)}{f'(1)} = 1 - \left(-\frac{3}{8}\right)$$
$$= 1.375$$

Exercise 10C

In each of Questions **1** to **6**, use the Newton–Raphson method once to find an approximation to a root of the given equation. Start with the initial value given, and give your answer to three significant figures.

1 Equation $x^3 + x^2 - 7x = 0$; initial value $x = 2$.

2 Equation $x^3 - 6x + 1 = 0$; initial value $x = 3$.

3 Equation $x^4 - 3x^2 - 3 = 0$; initial value $x = 2$.

4 Equation $3x^4 - 4x^3 + 8x - 1 = 0$; initial value $x = -1$.

5 Equation $x^3 + 3x - 7 = 0$; initial value $x = 1$.

FP1

10.5 Step-by-step solution of differential equations

If you are told that $\dfrac{dy}{dx} = 3x^2$ and are asked to find y, you are being invited to solve a **differential equation**.

A differential equation is an equation which contains a differential expression.

The solution of $\dfrac{dy}{dx} = 3x^2$ is $y = x^3 + c$ because $\dfrac{d}{dx}(x^3 + c) = 3x^2$ for all constants c.

To solve a differential equation, you need to know one point on the curve. Many differential equations cannot be solved exactly, but if you know one point on the curve, you can find approximations to nearby points. This is called a **step-by-step method**.

> Unless you know a point on the curve, you cannot find c, because, for example, both $y = x^3 + 1$ and $y = x^3 + 10$ are solutions of $\dfrac{dy}{dx} = 3x^2$.

The Euler formula

> The Euler formula is:
> $$y_{n+1} \approx y_n + h f(x_n)$$

This formula is used to find approximate solutions for a differential equation of the form:

$$\frac{dy}{dx} = f(x)$$

where f is a given function.

> Note that f is used for the gradient function of y, not for y itself.

If you know the coordinates of a point (x_n, y_n) on a curve and you are told the value of the 'step length' h, then you can work out an approximation to y_{n+1}, which is the value of y at $x_n + h$.

For examination purposes, you need to know how to use this formula, but it may be helpful to know how to derive it.

In the diagram on the right, $P(x_0, y_0)$ is a point on the curve $y = f(x)$ and $Q(x_1, y_1)$ is another point on the curve close to P, where $x_1 - x_0 = h$ and h is small.

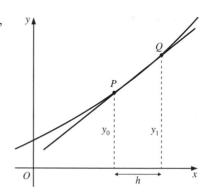

The gradient of the chord PQ is approximately the same as the gradient of the tangent at P. Hence,

$$\text{Gradient of } PQ = \frac{y_1 - y_0}{h}$$

which gives

$$\text{Gradient of tangent at } P = f(x_0) \approx \frac{y_1 - y_0}{h}$$

In general, this gives: $f(x_n) \approx \dfrac{y_{n+1} - y_n}{h}$

$$\text{or} \quad y_{n+1} \approx y_n + hf(x_n)$$

This is the Euler formula.

> This technique is known as **Euler's method**, after Leonhard Euler (1707–83), the prolific Swiss mathematician.
> It is similar to the procedure described in Section 8.1 for differentiating from first principles.

FP1

Example 6

The variables x and y satisfy the differential equation $\dfrac{dy}{dx} = \ln x$, and $y = 5$ when $x = 2$. Use the Euler formula with step length 0.1 to find an approximation for the value of y when $x = 2.2$.
....

Use the formula $y_{n+1} \approx y_n + hf(x_n)$.

In this case, $x_n = 2$, and $y_n = 5$.

The step length h is 0.1.

The function f is the natural logarithm function, so $f(x_n) = \ln 2 \approx 0.693$.

Thus, the formula gives $y_{n+1} \approx 5 + (0.1)(0.693) = 5.0693$.

This is the approximate value of y when $x = 2.1$.

Use the formula once more to find a value for y when $x = 2.2$.

This time $x_n = 2.1$ and $y_n = 5.0693$.

The step length h is unchanged but $f(x_n) = \ln 2.1 \approx 0.742$.

So, the formula gives $y_{n+1} \approx 5.0693 + (0.1)(0.742) = 5.143$.

And this is the required approximate value for the value of y when $x = 2.2$.

> The natural logarithm ln x is not examined in the FP1 module.

Example 7

The variables x and y satisfy the differential equation $\dfrac{dy}{dx} = e^{\cos x}$,

and $y = 3$ when $x = 1$.

Use the Euler formula with step length 0.2 to find an approximation for the value of y when $x = 1.4$.

· ·

Start with $x_n = 1$ and $y_n = 3$. The step length h is 0.2.

$$f(x_n) = e^{\cos 1} \approx 1.717$$

So, the formula gives $y_{n+1} \approx 3 + (0.2)(1.717) \approx 3.343$.

This is the approximate value of y when $x = 1.2$.

Then, using $x_n = 1.2$, $y_n = 3.343$ and $f(x_n) = e^{\cos 1.2} \approx 1.437$, apply the formula to obtain the result

$$y_{n+1} \approx 3.343 + (0.2)(1.437)$$

$$\approx 3.631.$$

> The exponential constant e is not examined in the FP1 module.

> You need to set your calculator in radian mode and find the cosine of 1 radian, then apply the inverse of the natural logarithm function. These techniques are described in module C2.

FP1

Exercise 10D

· ·

In each of Questions **1** to **7**, use the Euler formula with the given step length to find an approximate value for y for the stated value of x. Give each answer to four significant figures.

> In Questions 1, 5 and 6, ensure that your calculator is in radian mode.

1 $\dfrac{dy}{dx} = x \cos x$, given that $y = 2$ when $x = 1$; $h = 0.1$; find y for $x = 1.2$.

2 $\dfrac{dy}{dx} = x^2 + x^3 - x^4$, given that $y = 1$ when $x = 0$; $h = 0.1$; find y for $x = 0.2$.

3 $\dfrac{dy}{dx} = x \log_{10} x$, given that $y = 3$ when $x = 2$; $h = 0.1$; find y for $x = 2.2$.

4 $\dfrac{dy}{dx} = 2^x$, given that $y = 3$ when $x = 1$; $h = 0.1$; find y for $x = 1.3$.

5 $\dfrac{dy}{dx} = x^2 \sin x$, given that $y = 2$ when $x = 1$; $h = 0.01$; find y for $x = 1.02$.

6 $\dfrac{dy}{dx} = 3^x \sin x$, given that $y = 1$ when $x = 0$; $h = 0.1$; find y for $x = 0.2$.

7 $\dfrac{dy}{dx} = x \ln x$, given that $y = 4$ when $x = 2$; $h = 0.01$; find y for $x = 2.02$.

Summary

You should know how to ...	Check out
1 Locate a root within an interval.	**1** Show that the equation $\sin x = 2 - x$ has a root between 1.1 and 1.2.
2 Use interval bisection to locate a root within a smaller interval.	**2** Given that the equation $2x^3 = 9 - 2x$ has a root between 1.4 and 1.5, use interval bisection to determine the root correct to one place of decimals.
3 Use linear interpolation to find an approximation to a root.	**3** Given that the equation $x^4 = 1 - x$ has a root between 0.7 and 0.8, use linear interpolation (once) to find an approximation to the root.
4 Use the Newton–Raphson method to find an approximation to a root.	**4** Use the Newton–Raphson method once, with starting value $x = 1.5$, to find an approximation to the root of the equation $x^2 - \sqrt{x} - 1 = 0$.
5 Use the Euler formula for numerical solution of differential equations.	**5** A curve passes through $(2, 3)$ and its gradient is $\log_{10} x$. Use the Euler formula twice, with step length 0.1, to estimate the value of y at $x = 2.2$.

FP1

Revision exercise 10

1 a) Show, by drawing a sketch, that the graphs of $y = x^3$ and $y = 1 - x$ have one point of intersection in the interval $0 < x < 1$.

   b) Show that the x-coordinate of this point lies between 0.6 and 0.7.

2 The function f is defined for all real values of x by

$$f(x) = (x^2 + 4)(2x - 1).$$

The curve $y = f(x)$ intersects the line $y = x$ at only one point B.

   a) Show that the x-coordinate of B satisfies the equation

$$2x^3 - x^2 + 7x - 4 = 0$$

   b) Show that this equation has a root between 0.56 and 0.57.   *(AQA, 2001)*

3 The diagram shows the graphs of

$$y = x^3 \text{ and } y = x + 1,$$

intersecting at the point P, which has x-coordinate α.

   a) Show that, at P:

$$x^3 - x - 1 = 0.$$

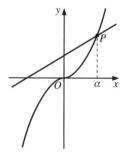

b) i)   Show that α lies in the interval between 1.2 and 1.4.

   ii) Use interval bisection **twice**, starting with the interval in part b) i), to find an interval of width 0.05 within which α must lie.

   iii) Hence give the value of α to one decimal place.

   (AQA, 2003)

4   The diagram shows the graph of

$$y = x\sin x \quad 0 \le x \le \frac{\pi}{2}.$$

a) Show, using a suitable diagram, that the equation

$$x\sin x = \cos x$$

has exactly one root in the interval $0 \le x \le \frac{\pi}{2}$.

b) Denoting this root by α, show that α is also a root of the equation f(x) = 0, where

$$f(x) = \tan x - \frac{1}{x}$$

c) Show that $f(0.8) = -0.220$ and find the value of f(0.9) to 3 decimal places.

d) Use linear interpolation once to estimate the value of α, giving your answer to 2 decimal places.

   (AQA, 2001)

FP1

5 a) Use logarithms to solve the equation $2^x = 7$, giving your answer to 3 significant figures.

   b) The equation

$$2^x = 7 - x$$

   has a single root, α.

   i) Show that α lies between 2.0 and 2.4.

   ii) Use the bisection method to find an interval of width 0.1 in which α lies.

   (AQA, 2003)

6   Whilst solving the equation f(x) = 0, where

$$f(x) = x^3 + 2x - 6,$$

a student correctly obtains $f(1) = -3$ and $f(2) = 6$. He concludes that the equation has a root lying between $x = 1$ and $x = 2$. He takes $x = 1$ as his first approximation to the root.

Find the next approximation to the root

a) using linear interpolation,

b) using the Newton–Raphson method.

   (AQA/NEAB, 2000)

7 A curve satisfies the differential equation
$$\frac{dy}{dx} = \frac{1}{1 + x^3}.$$
Starting at the point $(1, 0.5)$ on the curve, use a step-by-step method with a step length of 0.25 to estimate the value of y at $x = 1.5$, giving your answer to 2 decimal places.

(AQA, 2004)

8 A curve satisfies the differential equation $\frac{dy}{dx} = \sqrt{x^2 - 5}$.

Starting at the point $(3, 1)$ on the curve, use a step-by-step method with a step length of 0.5 to estimate the value of y at $x = 4$, giving your answer to 2 decimal places.

(AQA, 2003)

FP1

9 A curve satisfies the differential equation $\frac{dy}{dx} = \sqrt{9 - x^2}$.

Starting at the point $(0, 3)$ on the curve, use a step-by-step method with a step length of 0.5 to estimate the value of y at $x = 1$, giving your answer to 2 decimal places.

(AQA, 2001)

10 A curve passes through $(1, 1)$ and its gradient is 10^{x-2}.

Starting at the point $(1, 1)$ on the curve, use a step-by-step method with a step length of 0.01 to estimate the value of y at $x = 1.04$, giving your answer to five significant figures.

11 A curve passes through $(2, 2)$ and its gradient is $\cos x$, where x is in **radians**.

Starting at the point $(2, 2)$ on the curve, use a step-by-step method with a step length of 0.25 to estimate the value of y at $x = 3$. Show all your working to five significant figures and give your final answer to three significant figures.

11 Linear laws

This chapter will show you how to

✦ Reduce a relation to a linear law
✦ Use logarithms for relations involving powers or exponentials

Before you start

You should know how to ...	Check in
1 Plot points on a graph and fit a straight line to the points where possible.	**1** Plot the points $(1, 3)$, $(2, 2.4)$, $(3, 2)$ and $(5, 1)$, and draw a straight line passing through three of the points and close to the fourth.
2 Find the gradient and y-intercept of a linear graph.	**2** Find the gradient and the y-intercept of the graph drawn in Question **1**.
3 Interpret the gradient and y-intercept using the equation $y = mx + c$.	**3** Write down the equation of the line drawn in Question **1**.
4 Estimate other values of x or y using interpolation or extrapolation.	**4** Use your linear graph drawn in Question **1** to estimate: a) the value of y corresponding to $x = 4$ (interpolation) b) the value of x corresponding to $y = 0$ (extrapolation)
5 Use the laws of logarithms.	**5** a) Express $\log_{10}(ab)$ in terms of $\log_{10} a$ and $\log_{10} b$. b) Express $\log_{10}(a^n)$ in terms of n and $\log_{10} a$.

Links to Core modules

The laws of logarithms are in module C2. However, you do not need to know about the exponential constant e, or natural logarithms, for the FP1 module examination.

11.1 Reducing a relation to a linear law

Liz is carrying out an experiment using an electrical circuit. She knows that the voltage, V volts, should be proportional to the current, I amps. Her results are shown in the table.

I	V
0	0
2	15.9
4	31.8
8	63.9
10	80.8

Plotting these values on a graph, Liz can draw a straight line through the points.

From this she can:

◆ Confirm that the voltage is proportional to the current.

◆ Find the gradient of the line.

◆ Find the value of the voltage when $I = 6$, which she forgot to write down.

From the straight line you can see that the missing value of V is 48.

Notice that the points do not lie exactly on the straight line, but within the limitations of a scientific experiment, the results agree with the proposed equation.

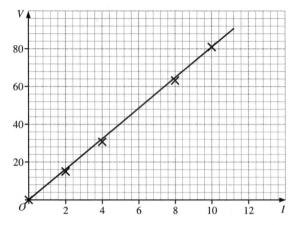

A similar experiment is performed. This time the voltage is kept constant and the resistance, R ohms, is measured as the current, I amps, is varied.

The results are shown in the table.

I	R
2	9.9
4	5.0
8	2.4
10	2.1

When these values are plotted on a graph of R against I, the points appear to lie on a curve in which R is inversely proportional to I. With a curve instead of a straight line, it is difficult to check the accuracy of the data or to estimate a realistic value of R when $I = 6$.

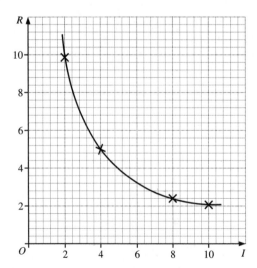

✦ A straight line can be obtained by plotting I against $\frac{1}{R}$
(or R against $\frac{1}{I}$). To help plot this, add a column, showing values
of $\frac{1}{R}$, to the table as shown.

I	R	$\frac{1}{R}$
2	9.9	0.101
4	5.0	0.2
8	2.4	0.417
10	2.1	0.476

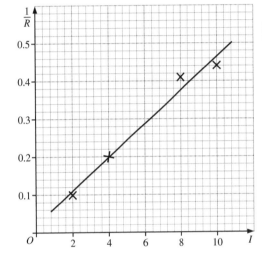

FP1

Plotting I against $\frac{1}{R}$ enables you to draw a straight line
through the points (remembering that experimental
results will not be exact).

Now you can:

✦ Confirm that I is proportional to $\frac{1}{R}$,
that is, the current is inversely proportional to the
resistance.

✦ Find the gradient of the line, which is 0.05 or $\frac{1}{20}$,
hence: $\frac{1}{R} = \frac{1}{20}I$

or $IR = 20$,

✦ Find the value of R when $I = 6$.

When $I = 6$, the value of $\frac{1}{R}$ is 0.3, which gives $R = 3.33$.

This technique can be applied to more complicated equations, and the
resulting straight-line graph will not necessarily pass through the origin.

For example, if you think that y^2 might be a linear function of x^3, you
would expect to find an equation of the form $y^2 = ax^3 + b$, where a and
b are constants.

Plotting y^2 against x^3 will produce a straight line with gradient a and
intercept b if this linear relationship exists.

> Suitable manipulation of variables enables you to make many
> equations into straight-line graphs.

For example, the equation $y^3 = ax^5 + bx^2$ can be made linear by
dividing by x^2, so that:

$$\frac{y^3}{x^2} = ax^3 + b$$

and then plotting $\frac{y^3}{x^2}$ against x^3.

Example 1

In an experiment these results are obtained:

x	y
1	1.34
2	1.53
3	1.78
4	2.05
5	2.29

It is believed that a formula such as $y^3 = ax^2 + b$ connects x and y.

Test to find if the results do satisfy such a formula and find the values of a and b.

To reduce the equation $y^3 = ax^2 + b$ to a linear law, you need to plot y^3 against x^2.

Draw up a new table:

x	y	x^2	y^3
1	1.34	1	2.406
2	1.53	4	3.582
3	1.78	9	5.640
4	2.05	16	8.615
5	2.29	25	12.01

Plotting y^3 against x^2 gives this graph.

The gradient of the line is $\dfrac{8}{20} = 0.4$.

The intercept with the y^3-axis is 2.

Thus, the results do satisfy an equation of the form $y^3 = ax^2 + b$, where $a = 0.4$ and $b = 2$, that is:

$$y^3 = 0.4x^2 + 2:$$

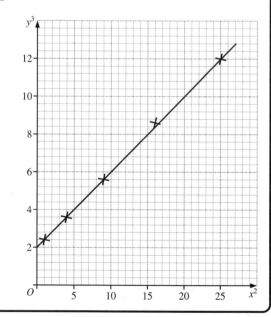

Exercise 11A

1 In an experiment, a ball is rolled down a slope. The distance of the ball from a marked point, x metres, is recorded at time t seconds from the start of the motion. The following results are obtained:

t	0	1	2	3	4
x	5	9.1	20.9	40.9	69.1

It is believed that a formula such as $x = at^2 + b$ connects x and t.

Test to find if the results do satisfy such a formula, and find the values of a and b.

2 In an experiment examining Boyle's law these results are obtained:

P	10	5	2	1	0.5
V	0.95	1.95	4.95	9.9	19.5

It is believed that a formula such as $PV = k$ connects V and P.

Test to find if the results do satisfy such a formula, and find the value of k.

3 In an experiment these results are obtained:

x	1	2	3	4	5
y	9.0	23.1	60.9	135	257

It is believed that a formula such as $y = ax^3 + b$ connects y and x.

Test to find if the results do satisfy such a formula, and find the values of a and b.

4 In an experiment the following results are obtained:

x	1	2	3	4	5
y	5.0	5.8	6.4	7.0	7.5

It is believed that a formula such as $y = a\sqrt{x} + b$ connects y and x.

Test to find if the results do satisfy such a formula and find the values of a and b.

5 It is believed that a formula such as $\dfrac{1}{x} + \dfrac{1}{y} = k$ connects y and x.

x	1	2	3	4	5
y	1.0	0.66	0.60	0.57	0.56

Test to see if the results do satisfy such a formula and find the value of k.

FP1

6 In an experiment the following results are obtained:

t	1	2	3	4	5
x	2.8	7.2	13.0	19.8	27.6

It is believed that a formula such as $x^2 = at^3 + bt$ connects x and t.

Test to find if the results do satisfy such a formula and find the values of a and b.

7 In an experiment these results are obtained:

x	1	2	3	4	5
y	3.2	5.6	9.4	14.1	19.5

Test the results to see if they satisfy a formula such as $y^2 = ax^3 + bx$, and find the values of a and b.

11.2 Use of logarithms

Relationships of the form $y = ax^n$

> Two variables x and y, which are related by a law of the form $y = ax^n$, can be turned into a linear relation if you use logarithms.

Use the laws of logarithms (introduced in Module C2).

From $y = ax^n$, you take logarithms:

$$\log y = \log (ax^n)$$

$$= \log a + \log x^n$$

$$= \log a + n \log x$$

You could use logarithms to any base, but most calculators only give logarithms to base 10 or base e.

> Use $\log xy = \log x + \log y$.

> Use $\log (x^k) = k \log x$

> You do not need to know about base e for the FP1 module.

Example 2

The volume of water flowing over a weir, $V\,\text{m}^3$, was measured together with the depth, x m, of the water above the weir. It is believed that V and x are connected by an equation of the form $V = ax^b$. The results were as follows:

x	y
0.5	2.6
1	20
1.5	71
2	170
4	1350
5	2600

Use a linear graph to confirm this relationship, and find the values of a and b.

Find also the value of V when $x = 3$.

FP1

· ·

$V = ax^b$.

Taking logarithms to base 10:

$\log_{10} V = \log_{10} a + b \log_{10} x$

Adding columns for $\log_{10} V$ and $\log_{10} x$ to the above table gives:

x	V	$\log_{10} x$	$\log_{10} V$
0.5	2.6	−0.301	0.415
1	20	0	1.301
1.5	71	0.176	1.851
2	170	0.301	2.230
4	1350	0.602	3.130
5	2600	0.699	3.415

Plotting values of $\log_{10} V$ against $\log_{10} x$ gives this graph.

The graph shows that the relationship is linear.

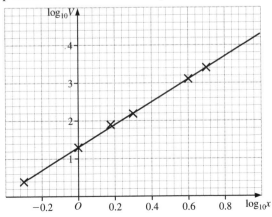

The gradient b is $\dfrac{1.50}{0.5} = 3$.

The intercept on the $\log_{10} V$ axis is 1.3.

So $\log_{10} a = 1.3$, hence:

$\qquad = 10^{1.3}$

$\qquad = 19.95 \ldots$

So, the relationship between V and x is approximately $V = 20x^3$.

When $x = 3$, $V = 540$.

> $\log_{10} x$ is zero at the $\log_{10} V$ intercept.

> When the depth of water is 3 metres, the volume flowing is 540 cubic metres.

Relationships of the form $y = ab^x$

Two variables x and y which are related by a law of the form $y = ab^x$ can be formed into a linear relation if you use logarithms.

Taking logarithms,

$$\log y = \log (ab^x)$$
$$= \log a + \log b^x$$
$$= \log a + x \log b$$

Once again you can use either base 10 or base e logarithms, that is:

$$\log_{10} y = \log_{10} a + x \log_{10} b \text{ or}$$
$$\log_e y = \log_e a + x \log_e b$$

Example 3

In an experiment lasting 11 days, the mass of a radioactive substance, M grams, was recorded on six different days. It is expected that M and the number of days, t, are connected by an equation of the form $M = ab^t$.

The results are shown in the table.

t	M
1	160
2	130
3	102
5	66
7	42
10	21

Use a linear graph to confirm that the predicted relationship is valid and find the values of a and b, giving your answers to two significant figures.

Find the mass of the substance which is present:
a) at the beginning, and b) at the end, of the experiment.

· ·

$M = ab^t$

Taking logarithms to base e: $\ln M = \ln a + t \ln b$

Adding a column for $\ln M$ to the above table gives:

> Logarithms to base e (natural logarithms) are not examined in the FP1 module.

FP1

t	M	$\ln M$
1	160	5.08
2	130	4.87
3	102	4.62
5	66	4.19
7	42	3.73
10	21	3.05

Plotting values of $\ln M$ against t gives the graph shown.

The gradient, $\ln b$, is $\dfrac{-1.14}{5} = -0.228$. Hence:

$b = e^{-0.228}$

$\quad = 0.796 \ldots$

$b = 0.80 \ (2 \text{ sf})$

The intercept on the $\ln M$ axis is

$\ln a = 5.30$

$\quad a = e^{5.30}$

$\quad a = 200 \ (2 \text{ sf})$

$M = ab^t$

$M = 200 \, (0.8)^t$

When $t = 0$, $M = 200$.

The initial mass is 200 grams.

When $t = 11$, $M = 200 \, (0.8)^{11} = 17$

After 11 days the mass is 17 grams.

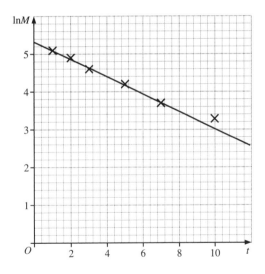

> $0.8^0 = 1$

> Extending the linear model to $t = 0$ and to $t = 11$ are examples of **extrapolation**, or predicting beyond the range of available data. Extrapolation can be unreliable.

Exercise 11B

1 In an experiment the mass, m grams, of bacteria present at time t days after the start of the experiment was recorded for certain values of t. The results are shown in the table.

t	1	2	3	4	5
m	24.0	28.9	34.5	41.5	49.8

It is believed that a formula such as $m = ab^t$ connects m and t.

a) i) Use a linear graph to confirm this relationship.

  ii) Find the values of a and b.

b) Write down the initial mass.

c) Predict the mass after 60 days.
    Comment on the reliability of your prediction.

2 In an experiment into radioactive decay, the mass, m grams, of a substance was recorded at time t hours after the start of the experiment. The results are shown in the table.

t	0	1	2	3	4
m	27.1	24.3		19.7	17.7

It is believed that a formula such as $m = ab^t$ connects m and t.

a) i) Use a linear graph to confirm this relationship.

  ii) Find the values of a and b.

b) Find the value of m corresponding to $t = 2$.

c) Predict the mass after seven hours.

3 The table shows the results of an experiment involving the quantities x and y.

x	1	2	3	4
y	6.9	19.8	36.4	55.9

It is believed that x and y are connected by a formula such as $y = ax^b$.

a) Use a linear graph to confirm this relationship.

b) Find the values of a and b.

FP1

4 The table shows the results of an experiment involving the quantities x and y.

x	1	2	3.5	4.9
y	5.09	17.8	48.6	89.1

It is believed that x and y are connected by a formula such as $y = ax^b$.

a) Use a linear graph to confirm this relationship.

b) Find the values of a and b.

5 In an experiment the mass, m grams, of bacteria present at time t days after the start of the experiment was recorded for certain values of t. The results are shown in the table.

t	1	2	3	4	5
m	102	52.0	29.5	13.5	6.91

It is believed that a formula such as $m = ab^t$ connects x and t.

a) Assuming that such a relationship exists, find the initial mass.

b) It is believed that one of the recordings of mass was incorrect. Which was this recording, and what should the mass have been?

FP1

Summary

You should know how to ...	Check out
1 Reduce a relation to a linear law.	**1** Define variables X and Y so that the following relations take the form $Y = mX + c$: a) $\dfrac{1}{x} - \dfrac{1}{y} = k$ b) $y^2 = ax^3 + b$ c) $y^2 = ax^3 + bx$
2 Use logarithms for relations involving powers or exponentials.	**2** Use logarithms to reduce the following relations to linear form: a) $y = x^n$ (n constant) b) $y = k^x$ (k constant) c) $u = \dfrac{k}{v^n}$ (k and n constant)

Revision exercise 11

1 The variables x and y are related by the equation $y = mx^2$, where m is a constant.
Measurements of y for given values of x gave the following results.

x	2	3	4	5
y	3	7	13	20

a) Make a table of values of X and Y, where $X = x^2$ and $Y = y$.
b) Plot the points with coordinates (X, Y) on graph paper.
c) Draw a straight line through the origin to illustrate the relationship between the variables.
d) Estimate the gradient of your linear graph.
e) Comment on the significance of the graph passing through the origin, and estimate the value of the constant m.

2 The points (x, y) are known to lie on a circle with equation $x^2 + y^2 = r^2$.

Some pairs of approximate coordinates are shown in the table.

x	1.2	2.1	2.4	3.1	3.6
y	3.8	3.4	3.2	2.5	1.8

a) Make a table of values of X and Y, where $X = x^2$ and $Y = y^2$.
b) Plot the points with coordinates (X, Y) on graph paper.
c) Draw a straight line to illustrate the relationship between X and Y.
d) Estimate the Y-intercept of your linear graph, and hence estimate the radius of the circle on which the points (x, y) lie.
e) Comment on the significance of the gradient of your linear graph.

3 The variables T and L satisfy a relationship of the form $T = aL^b$, where a and b are constants.
Measurements of T for given values of L gave the following results.

L	2	3	4	5	6
T	5.62	6.94	8.03	8.98	9.97

a) Express $\log_{10} T$ in terms of a, b and $\log_{10} L$.
b) Plot $\log_{10} T$ against $\log_{10} L$ on graph paper.
c) Draw a suitable straight line to illustrate the relationship between the data.
d) Use your line to estimate:
   i) the value of L when $T = 8.50$, giving your answer to two significant figures,
   ii) the values of a and b, giving your answers to two significant figures.   *(AQA, 2001)*

4 The diagram shows a straight-line graph of $\log_{10} y$ against $\log_{10} x$ passing through the points $(0, 2)$ and $(3, 6)$.

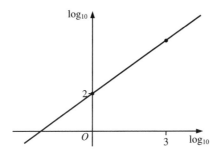

Express y in the form ax^b, where a and b are constants to be determined to three significant figures.

(AQA/NEAB, 2000)

FP1

5 The variables Q and x satisfy a relationship of the form $Q = ax^b$, where a and b are constants. Measurements of Q for given values of x gave the following results.

x	0.4	0.5	0.6	0.7	0.8
Q	1.72	3.02	4.74	6.98	9.73

a) Express $\log_{10} Q$ in terms of $\log_{10} a$, b and $\log_{10} x$.

b) i) Copy and complete the following table.

x	0.4	0.5	0.6	0.7	0.8
$\log_{10} x$					
Q	1.72	3.02	4.74	6.98	9.73
$\log_{10} Q$					

ii) Plot $\log_{10} Q$ against $\log_{10} x$ on graph paper.
iii) Draw a suitable line to illustrate the relationship between the data.

c) Use your line to estimate:
i) the value of Q when $x = 0.54$, giving your answer to two significant figures,
ii) the values of a and b, giving your answers to two significant figures.

(AQA, 2004)

6 A mathematical model is required to estimate the number, N, of a certain strain of bacteria in a test tube at time t hours after a certain instant.

After values of $\log_{10} N$ are plotted against t, a straight line graph can be drawn through the points as shown below.

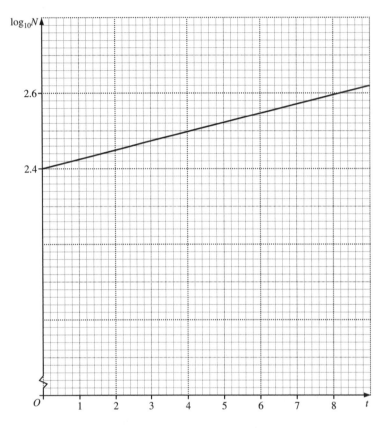

FP1

a) Use the graph to estimate the number of bacteria when $t = 5$.

b) The graph would suggest that N and t are related by an equation of the form

   $$N = a \times b^t$$

   where a and b are constants.
   i) Express $\log_{10} N$ in terms of $\log_{10} a$, $\log_{10} b$ and t.
   ii) Use the graph to determine the values of a and b, giving your answers to 3 significant figures.

c) Suggest why the model $N = a \times b^t$ is likely to give an overestimate of the number of bacteria in the test tube for large values of t.

(AQA, 2003)

7 The power, P watts, dissipated by a resistor was measured for varying values of current, I amperes, flowing in the resistor. The resulting values of $\log_{10} I$ and $\log_{10} P$ are given in the table below.

$\log_{10} I$	0.336	0.742	1.28	1.55	1.76
$\log_{10} P$	3.76	4.57	5.65	6.19	6.61

It is thought that the relationship between I and P can be modelled by the equation

$$P = kI^n,$$

where k and n are constants.

a) Express $\log_{10} P$ in terms of $\log_{10} k$, $\log_{10} I$ and n.

b) Use the data in the table to draw a suitable linear graph to illustrate this model.

c) Explain why your graph is consistent with the suggested model.

d) Use your graph to obtain approximate values, to two significant figures, for

i) n,

ii) k.

(AQA/NEAB, 2001)

FP1

8 In an experiment, a radio-controlled car was driven along a straight track starting from a fixed point O. The displacement, x metres, of the car after time t seconds was observed and the following results were recorded.

t	3	4	5	6	8	10
x	15.9	22.4	29.0	36.6	53.6	74.0

a) i) From these results, calculate the six values of $\dfrac{x}{t}$.

ii) On a sheet of graph paper, plot the corresponding six points with coordinates $\left(t, \dfrac{x}{t}\right)$.

b) Explain how these points support the constant acceleration model

$$x = ut + \tfrac{1}{2}at^2,$$

where u m s^{-1} denotes the speed at O and a m s^{-2} denotes the acceleration.

c) Use your diagram to estimate values for u and a, giving your answers to two significant figures.

(AQA/NEAB, 2001)

9 The volume, y m³, of water flowing per second over a weir is measured when the difference of levels is x metres.

Corresponding values of the two variables x and y are given in the table.

x	1.1	1.3	1.5	1.7	1.9
y	5.3	7.9	11.1	15.0	19.6

It is thought that the relationship between x and y can be modelled by the equation

$$y = ax^n$$

where a and n are constants.

a) Express $\log_{10} y$ in terms of x, a and n.

b) Use the data in the table to draw a suitable linear graph to illustrate this model.

c) Explain why your graph is consistent with the suggested model.

d) Use your graph to obtain approximate values, to two significant figures, for

   i) a,

   ii) n.

(AQA/NEAB, 1999)

FP1

MFP1 Practice Paper

90 minutes   75 marks   You may use a calculator

1 The matrices **A**, **B** and **C** are given by

$$\mathbf{A} = \begin{bmatrix} 1 & 2 \\ 3 & 4 \end{bmatrix}, \ \mathbf{B} = \begin{bmatrix} 2 & 1 \\ -2 & 5 \end{bmatrix}, \ \mathbf{C} = \begin{bmatrix} 5 & -1 \\ 2 & 2 \end{bmatrix}$$

a) Calculate the matrices

   i) **BC**,   *(2 marks)*

   ii) **ABC**.   *(2 marks)*

b) Describe the geometrical transformation represented by the
matrix **BC**.   *(2 marks)*

FP1

2 The complex number z is equal to $x + iy$, where x and y are real
numbers.

a) Given that z^* is the conjugate of z, write $z^* - i$ in the form
$p + iq$.   *(2 marks)*

b) Given that

$$3(z + i) = i(z^* - i)$$

find the value of the complex number z.   *(4 marks)*

3 The roots of the quadratic equation

$$2x^2 - x - 2 = 0$$

are α and β.

a) Write down the values of $\alpha + \beta$ and $\alpha\beta$.   *(2 marks)*

b) Hence, without solving the equation, find the numerical
value of

   i) $\alpha^2 + \beta^2$,   *(2 marks)*

   ii) $(\alpha - \beta)^2$.   *(3 marks)*

4 The function f is defined by $f(x) = (x - 1)^2$.

a) Find $f(3 + h) - f(3)$, giving your answer in the form $ah + bh^2$.   *(4 marks)*

b) Use your answer to part a) to find the value of $f'(3)$.   *(2 marks)*

5 a) Use the formula

$$\sum_{r=1}^{n} r^2 = \tfrac{1}{6}n(n+1)(2n+1)$$

to show that

$$\sum_{r=1}^{n} (r^2 - 1) = \tfrac{1}{6}n(n-1)(2n+5)$$ *(3 marks)*

b) Hence show that $14^2 + 15^2 + 16^2 + 17^2 - 4$ is divisible by 13.   *(4 marks)*

FP1

6 A curve passes through the point (2, 3) and at every point on the curve

$$\frac{dy}{dx} = \log_{10} x$$

Use a step-by-step method starting at the point (2, 3), and a step length $h = 0.2$, to estimate the value of y when $x = 2.6$. Show your calculations to 4 decimal places and give your final answer to 2 decimal places.   *(7 marks)*

7 a) Write down the exact values of $\sin\dfrac{\pi}{6}$, $\cos\dfrac{\pi}{6}$ and $\tan\dfrac{\pi}{6}$.   *(3 marks)*

b) Find the general solution of the equation

$$2\cos\left(\theta - \frac{\pi}{3}\right) = \sqrt{3}$$

Give all solutions in terms of π.   *(6 marks)*

8 A curve has equation

$$y = \frac{x^2 - 4x + 3}{x^2 - x - 2}$$

a) Show that the curve has one horizontal and two vertical asymptotes, and find the equations of these three asymptotes.   *(4 marks)*

b) Find the coordinates of the three points where the curve intersects the coordinate axes.   *(4 marks)*

c) Show that the curve intersects the line $y = 1$ at only one point, and find the coordinates of this point.   *(3 marks)*

d) Sketch the curve, showing all the asymptotes and all the points of intersection with the coordinate axes and with the line $y = 1$.   *(4 marks)*

9 a) Sketch the ellipse C which has equation

$$\frac{x^2}{4} + y^2 = 1$$

showing the coordinates of the points where the ellipse
intersects the axes.   *(3 marks)*

b) Describe a geometrical transformation which would transform
the unit circle

$$x^2 + y^2 = 1$$

into the ellipse C.   *(2 marks)*

c) Show that, if the line L which has equation

$$x + 2y = 2\sqrt{2}$$

intersects the ellipse C, then the x-coordinates of the points of
intersection must satisfy the quadratic equation

$$x^2 - 2\sqrt{2}x + 2 = 0$$ *(4 marks)*

d) By considering the discriminant of this quadratic equation,
or otherwise, determine whether L is

a tangent to C,

a line intersecting C in two distinct points, or

a line which does not intersect C.   *(3 marks)*

FP1

Answers

Chapter 1

Check in

1 a) $-1, -3$ b) $-2, -2$ c) $1, \frac{1}{2}$ d) $1, \frac{1}{3}$ **2** a) $x^2 + 2x + 1$ b) $4x^2 - 12x + 9$ c) $x^2 + 2 + \dfrac{1}{x^2}$

5 a) $\dfrac{a+b}{ab}$ b) $\dfrac{b-a}{ab}$ c) $\dfrac{1}{ab}$ d) $\dfrac{a^2+b^2}{ab}$ e) $\dfrac{a}{c}$ f) $\dfrac{a^2-b^2}{abc}$

Exercise 1A

1 a) $-3, -7$ b) $11, 5$ c) $-5, -4$ d) $-\frac{11}{3}, \frac{2}{3}$ e) $-2, -5$ f) $-2, -\frac{7}{2}$ **2** a) $x^2 - 7x + 12 = 0$ b) $x^2 + 3x + 2 = 0$

2 c) $x^2 + 2x - 4 = 0$ d) $x^2 + 5x - 11 = 0$ **3** a) -3 b) $\frac{1}{2}$ **4** a) -2 b) $\frac{1}{3}$ **5** $3x^2 - 16x + 20 = 0$

Exercise 1B

1 a) $2x^2 + 35x - 392 = 0$ b) $8x^2 - 5x - 2 = 0$ c) $4x^2 - 57x + 64 = 0$ d) $8x^2 + 365x - 512 = 0$ e) $16x^2 + 190x - 1889 = 0$

1 f) $8x^2 + 135x - 114 = 0$ **2** a) $3x^2 + 5x - 6 = 0$ b) $3x^2 - 20x - 96 = 0$ c) $9x^2 - 61x + 36 = 0$

2 d) $27x^2 - 395x - 216 = 0$ e) $6x^2 - 5x - 173 = 0$

Exercise 1C

1 $x^2 + 14x + 44 = 0$ **2** $x^2 - 45x + 63 = 0$ **3** $2x^2 - x + 16 = 0$ **4** $3x^2 + 23x + 55 = 0$

5 a) $x^2 + 7x + 6 = 0$ b) $6x^2 + 7x + 1 = 0$ c) $2x^2 - x - 3 = 0$ d) $x^2 + 5x = 0$

6 a) $3x^2 + 36x - 32 = 0$ b) $6x^2 + 9x - 1 = 0$ c) $3x^2 + 27x + 52 = 0$ d) $x^2 + 13x + 16 = 0$ **7** $x^2 + 10x + 75 = 0$

Check out

1 a) $6, 8$ b) $-\frac{5}{2}, -\frac{1}{2}$ **2** a) $\frac{1}{9}$ b) $\frac{13}{9}$ c) $\frac{4}{9}$ **3** a) $x^2 + 12x + 18 = 0$ b) $2x^2 + 12x + 9 = 0$

Revision exercise 1

1 a) $5, 6; 2, 3$ b) $-1, -6; 2, -3$ c) $6, 0; 0, 6$ d) $0, -9; 3, -3$ **2** a) $x^2 - 3x - 2 = 0$ b) $x^2 + 6x - 8 = 0$ c) $2x^2 - 3x - 1 = 0$

3 a) i) 22 ii) $-\frac{1}{3}$ b) $3x^2 - 74x - 1 = 0$ **4** a) i) $\frac{1}{2}$ ii) 3 b) i) 2 ii) 6 c) $x^2 - 6x + 2 = 0$

5 a) ii) 18 b) ii) 47 c) $x^2 - 15x - 45 = 0$ **6** a) $-3, -2$ b) i) $1\frac{3}{4}$ ii) $-\frac{17}{4}$ c) $4x^2 + 51x - 17 = 0$ **7** $x^2 - 7x + 9 = 0$

8 a) i) $\frac{22}{9}$ ii) $-\frac{22}{27}$ b) $81x^2 + 66x + 1 = 0$ **9** a) 2 b) i) $-p$ ii) $p^2 - 4$ c) $3, -3$

10 a) $-7 - p, p$ b) $(7+p)^2 - 2p$ c) ii) $-4, -6$

Chapter 2

Check in

1 a) $(x+1)(x+2)(2x+3)$ b) $3(x+1)(x+2)$ c) $8x^2(2x+3)$ d) $3(x+1)^2(x-2)^2$ **2** a) $z = 2(x+1)(4x+3)$

2 b) $z = x^2(x^4 + 2)$ c) $z = x^2$

Exercise 2A

1 a) 55 b) 325 c) 1425 d) 98 e) 270 f) 484 **2** $p = 18, q = 42$

Exercise 2B

1 a) 1240 b) 5740 c) $16\,600$ **2** a) $\dfrac{n}{2}(n+1)(2n+1)$ b) $\dfrac{n}{3}(4n^2 + 9n + 8)$ c) $\dfrac{n}{3}(5n^2 + 12n - 11)$

3 $11\,480$ **4** $25\,330$

Exercise 2C

1 $\dfrac{2n}{3}(n+1)(n+2)$ **2** $\dfrac{n}{2}(n+1)(n^2 + n + 1)$ **3** $\dfrac{n}{3}(n^2 - 7)$ **4** $\dfrac{n}{6}(4n^2 + 33n - 1)$ **5** $n(2n^3 - 8n^2 + 11n - 6)$

6 $44\,485$ **7** $N^2(2N+1)^2; -N^2(4N+3)$

Check out

1 a) $\sum_{r=100}^{200} r$ b) $\sum_{r=10}^{20} r^2$ c) $\sum_{r=1}^{25} (2r)^3$ **2** a) $\frac{1}{12}n(n+1)(3n^2+7n+2)$ b) $\frac{1}{12}n(n-1)(3n^2+n-2)$ c) $\frac{1}{3}n(2n+1)(6n^2-n-1)$

3 a) 2870 b) 2585 c) 845 000

Revision exercise 2

1 $\frac{1}{6}n(n+1)(2n+1), \frac{1}{6}n(2n^2+3n+7), \frac{1}{6}n(2n^2+3n-29)$ **2** $\frac{1}{4}n^2(n+1)^2, \frac{1}{4}n(n+1)(n^2+n+2), \frac{1}{4}n(n+1)(n^2+n-10)$

3 $\frac{1}{3}n(2n+1)(4n+1), \frac{9}{4}n^2(3n+1)^2$ **4** a) i) 25 502 500 ii) 23 876 875 b) 3775 c) 0   **5** a) 45 150 b) 9 090 200

7 a) $\frac{1}{3}n(n+1)(2n+1)$ b) $\frac{1}{2}n(n+1)$ **8** b) 332 999 700   **10** 3, −1

Chapter 3

Check in

1 a) 44 b) 126   **2** a) 4 b) −11

Exercise 3A

1 a) $\begin{pmatrix} 5 \\ -1 \end{pmatrix}$ b) $\begin{pmatrix} 9 \\ 12 \end{pmatrix}$ c) $\begin{pmatrix} 4 \\ -10 \end{pmatrix}$ d) $\begin{pmatrix} 25 \\ 18 \end{pmatrix}$ e) $\begin{pmatrix} 1 \\ 32 \end{pmatrix}$ **2** a) $\begin{pmatrix} 4 & 8 \\ -2 & 6 \end{pmatrix}$ b) $\begin{pmatrix} 12 & -21 \\ 9 & 3 \end{pmatrix}$ c) $\begin{pmatrix} 16 & -13 \\ 7 & 9 \end{pmatrix}$ d) $\begin{pmatrix} -14 & 47 \\ -18 & 4 \end{pmatrix}$

3 a) $\begin{pmatrix} 10 & 40 \\ 20 & -10 \end{pmatrix}$ b) $\begin{pmatrix} 8 & 15 \\ 11 & 1 \end{pmatrix}$ c) $\begin{pmatrix} 2 & 25 \\ 9 & -11 \end{pmatrix}$ d) $\begin{pmatrix} -12 & 37 \\ 1 & -33 \end{pmatrix}$

Exercise 3B

1 a) $\begin{pmatrix} 13 \\ 45 \end{pmatrix}$ b) $\begin{pmatrix} 3 \\ 14 \end{pmatrix}$ c) $\begin{pmatrix} 4 \\ 17 \end{pmatrix}$ **2** a) $\begin{pmatrix} 4 \\ -9 \end{pmatrix}$ b) $\begin{pmatrix} -23 \\ -8 \end{pmatrix}$ c) $\begin{pmatrix} -7 \\ -9 \end{pmatrix}$

Exercise 3C

1 a) $\begin{pmatrix} 16 & 14 \\ 38 & 43 \end{pmatrix}$ b) $\begin{pmatrix} 11 & 12 \\ 31 & 48 \end{pmatrix}$ **2** a) i) $\begin{pmatrix} -1 & 14 \\ 7 & -8 \end{pmatrix}$ ii) $\begin{pmatrix} -1 & 14 \\ 7 & -8 \end{pmatrix}$ iii) $\begin{pmatrix} 9 & -16 \\ -8 & 17 \end{pmatrix}$ b) $PQ = QP$

3 a) $\begin{pmatrix} 16 & 9 \\ 18 & -3 \end{pmatrix}$ b) $\begin{pmatrix} 3 & 16 \\ 15 & 10 \end{pmatrix}$ **4** a) $\begin{pmatrix} 0 & -10 \\ 12 & 0 \end{pmatrix}$ b) $\begin{pmatrix} 0 & -20 \\ 6 & 0 \end{pmatrix}$ **5** a) $\begin{pmatrix} 4 & 0 \\ 0 & 4 \end{pmatrix}$ b) $\begin{pmatrix} 8 & 0 \\ 0 & 8 \end{pmatrix}$ c) $\begin{pmatrix} 16 & 0 \\ 0 & 16 \end{pmatrix}$

6 a) $\begin{pmatrix} 0 & 3 \\ -3 & 0 \end{pmatrix}$ b) $\begin{pmatrix} 0 & 3 \\ -3 & 0 \end{pmatrix}$ c) $\begin{pmatrix} -\frac{1}{2} & \frac{\sqrt{3}}{2} \\ -\frac{\sqrt{3}}{2} & -\frac{1}{2} \end{pmatrix}$ d) $\begin{pmatrix} -1 & 0 \\ 0 & -1 \end{pmatrix}$ e) $\begin{pmatrix} 1 & 0 \\ 0 & 1 \end{pmatrix}$ f) $\begin{pmatrix} -\frac{3}{2} & \frac{3\sqrt{3}}{2} \\ -\frac{3\sqrt{3}}{2} & -\frac{3}{2} \end{pmatrix}$

6 g) $\begin{pmatrix} \frac{3}{2} & \frac{3\sqrt{3}}{2} \\ -\frac{3\sqrt{3}}{2} & \frac{3}{2} \end{pmatrix}$ h) $\begin{pmatrix} -\frac{\sqrt{3}}{2} & \frac{1}{2} \\ -\frac{1}{2} & -\frac{\sqrt{3}}{2} \end{pmatrix}$ **7** a) M b) M c) 0 d) 0

Check out

1 a) $\begin{pmatrix} 1 & 0 \\ 0 & 1 \end{pmatrix}$ b) $\begin{pmatrix} 17 \\ 39 \end{pmatrix}$ c) $\begin{pmatrix} 1 & 3 \\ 3 & 10 \end{pmatrix}$ **2** a) A b) A c) O d) O

Revision exercise 3

1 a) $\begin{pmatrix} 0 & 3 \\ 0 & 1 \end{pmatrix}$ b) $\begin{pmatrix} 0 & 6 \\ 0 & 2 \end{pmatrix}$ c) $\begin{pmatrix} 0 & 6 \\ 0 & 2 \end{pmatrix}$ **2** a) $\begin{pmatrix} 0 \\ 16 \end{pmatrix}$ b) $\begin{pmatrix} 0 \\ 32 \end{pmatrix}$ c) $\begin{pmatrix} 0 \\ 32 \end{pmatrix}$ **3** a) $\begin{pmatrix} 11 & 25 \\ 2 & 10 \end{pmatrix}$ b) $\begin{pmatrix} -6 & 6 \\ 18 & 12 \end{pmatrix}$ c) $\begin{pmatrix} 5 & 31 \\ 20 & 22 \end{pmatrix}$

4 a) $\begin{pmatrix} 2 & 7 \\ -1 & 10 \end{pmatrix}$ b) $\begin{pmatrix} 2 & -2 \\ 3 & -6 \end{pmatrix}$ c) $\begin{pmatrix} 7 & -10 \\ 10 & -22 \end{pmatrix}$ d) $\begin{pmatrix} 7 & -10 \\ 10 & -22 \end{pmatrix}$ **5** a) $\begin{pmatrix} 2p & 2q \\ 3r & 3s \end{pmatrix}$ b) $\begin{pmatrix} 2p & 3q \\ 2r & 3s \end{pmatrix}$ **6** a) $\begin{pmatrix} 4 & 7 \\ -7 & 4 \end{pmatrix}$ b) $\begin{pmatrix} 4 & 7 \\ -7 & 4 \end{pmatrix}$

7 a) $\begin{pmatrix} 5 & 3 \\ 7 & 5 \end{pmatrix}, \begin{pmatrix} 4 & 4 \\ 5 & 6 \end{pmatrix}$ b) $(A+B)^2 = \begin{pmatrix} 19 & 15 \\ 25 & 24 \end{pmatrix}, A^2 + 2AB + B^2 = \begin{pmatrix} 20 & 14 \\ 27 & 23 \end{pmatrix}$ c) $(A+B)(A-B) = \begin{pmatrix} 3 & 5 \\ 3 & 8 \end{pmatrix}, A^2 - B^2 = \begin{pmatrix} 4 & 4 \\ 5 & 7 \end{pmatrix}$

8 $x = pr - qs, y = ps + qr$ **9** $x = pr + qs, y = -ps + qr$ **10** a) $x = pr - qs, y = ps + qr$ b) Equality if $q = 0$ or if $r = s = 0$

Chapter 4

Check in

1 a) $(-2, 1)$ b) $(1, -2)$ c) $(2, 1)$ d) $(1, 6)$ e) $(2, 4)$ **2** a) $0.342, 0.940$ b) $9.06, 4.23$ **3** a) $\begin{pmatrix} 1 \\ -2 \end{pmatrix}$ b) $\begin{pmatrix} 0 & -3 \\ 1 & 0 \end{pmatrix}$

Exercise 4A

1 $\begin{pmatrix} 4 & 0 \\ 0 & 4 \end{pmatrix}$ **2** $\begin{pmatrix} 1 & 0 \\ 0 & -1 \end{pmatrix}$ **3** $\begin{pmatrix} -1 & 0 \\ 0 & 1 \end{pmatrix}$ **4** $\begin{pmatrix} 0 & -1 \\ -1 & 0 \end{pmatrix}$ **5** $\begin{pmatrix} 1 & 0 \\ 0 & 2 \end{pmatrix}$ **6** $\begin{pmatrix} 3 & 0 \\ 0 & 1 \end{pmatrix}$ **7** $\begin{pmatrix} 0 & -1 \\ 1 & 0 \end{pmatrix}$

8 $\begin{pmatrix} \cos 20° & -\sin 20° \\ \sin 20° & \cos 20° \end{pmatrix}$ **9** $\begin{pmatrix} \cos 40° & \sin 40° \\ -\sin 40° & \cos 40° \end{pmatrix}$ **10** $\begin{pmatrix} \cos 40° & \sin 40° \\ \sin 40° & -\cos 40° \end{pmatrix}$ **11** $\begin{pmatrix} \cos 40° & -\sin 40° \\ -\sin 40° & -\cos 40° \end{pmatrix}$

Exercise 4B

1 $\begin{pmatrix} \frac{\sqrt{3}}{2} & -\frac{1}{2} \\ \frac{1}{2} & \frac{\sqrt{3}}{2} \end{pmatrix}$ **2** $\begin{pmatrix} \frac{1}{2} & \frac{\sqrt{3}}{2} \\ -\frac{\sqrt{3}}{2} & \frac{1}{2} \end{pmatrix}$ **3** $\begin{pmatrix} -\frac{1}{\sqrt{2}} & \frac{1}{\sqrt{2}} \\ -\frac{1}{\sqrt{2}} & -\frac{1}{\sqrt{2}} \end{pmatrix}$ **4** $\begin{pmatrix} \frac{1}{2} & \frac{\sqrt{3}}{2} \\ \frac{\sqrt{3}}{2} & -\frac{1}{2} \end{pmatrix}$ **5** $\begin{pmatrix} \frac{1}{2} & -\frac{\sqrt{3}}{2} \\ -\frac{\sqrt{3}}{2} & -\frac{1}{2} \end{pmatrix}$ **6** $\begin{pmatrix} \frac{1}{\sqrt{2}} & \frac{1}{\sqrt{2}} \\ \frac{1}{\sqrt{2}} & -\frac{1}{\sqrt{2}} \end{pmatrix}$

Exercise 4C

1 An enlargement scale factor 3 with the origin as the centre of enlargement.

2 An enlargement scale factor 2 and centre O combined with a rotation about O through 180°.

3 A stretch parallel to the x-axis with scale factor 2.

4 A stretch parallel to the y-axis with scale factor 6.

5 A rotation through 60° clockwise with the origin as the centre of the rotation.

6 A half turn about the origin.

7 A stretch parallel to the y-axis with scale factor 2.

8 A rotation through 45° clockwise with the origin as the centre of the rotation.

Exercise 4D

1 $\begin{pmatrix} 0 & -3 \\ 3 & 0 \end{pmatrix}$ **2** $\begin{pmatrix} 0 & 4 \\ -4 & 0 \end{pmatrix}$ **3** $\begin{pmatrix} 5 & 0 \\ 0 & 3 \end{pmatrix}$ **4** $\begin{pmatrix} 0 & 7 \\ 1 & 0 \end{pmatrix}$ **5** $\begin{pmatrix} 0 & 1 \\ 5 & 0 \end{pmatrix}$ **6** $\begin{pmatrix} 0 & k \\ k & 0 \end{pmatrix}$ **7** $\begin{pmatrix} 0 & 4 \\ 32 & 0 \end{pmatrix}$

Note that there are other possible solutions to questions 8–15.

8 A stretch parallel to the x-axis with scale factor -2, followed by a stretch parallel to the y-axis of scale factor 2.

9 A stretch parallel to the x-axis with scale factor 4, followed by a stretch parallel to the y-axis with scale factor -1.

10 A stretch parallel to the x-axis with scale factor 3, followed by a stretch parallel to the y-axis with scale factor 5.

11 A half turn about the origin followed by an enlargement centre O scale factor 3.

12 A reflection in the x-axis followed by an enlargement centre O scale factor 3.

13 A reflection in the x-axis followed by a rotation through 45° anti-clockwise with the origin as the centre of rotation.

14 A rotation through 60° clockwise with the origin as the centre of rotation followed by an enlargement, centre O scale factor 2.

15 A rotation through 60° clockwise with the origin as the centre of rotation.

Check out

1 $\frac{\sqrt{3}}{2}, \frac{1}{\sqrt{2}}, \sqrt{3}$ **2** a) $\begin{pmatrix} -\frac{1}{2} & -\frac{\sqrt{3}}{2} \\ \frac{\sqrt{3}}{2} & -\frac{1}{2} \end{pmatrix}$ b) $\begin{pmatrix} -1 & 0 \\ 0 & 1 \end{pmatrix}$ c) $\begin{pmatrix} 4 & 0 \\ 0 & 1 \end{pmatrix}$ d) $\begin{pmatrix} 10 & 0 \\ 0 & 10 \end{pmatrix}$

3 a) Rotation 180° about O

b) reflection in $y = x$ and enlargement scale factor 2 about O

c) stretch scale factor 4 in x direction and scale factor 3 in y direction.

4 a) $\begin{pmatrix} 0 & -1 \\ 1 & 0 \end{pmatrix}$ b) rotation through 90° anticlockwise about O

Revision exercise 4

1 a) (0, 0), (3, 0), (7, 2), (4, 2); shear

1 b) (0, 0), (3, 3√3), (3 − 2√3, 2 + 3√3), (−2√3, 2); rotation about O through 60° anticlockwise and enlargement with centre O and scale factor 2

1 c) (0, 0), (3, 3√3), (3 + 2√3, −2 + 3√3), (2√3, −2); reflection in line $y = \dfrac{1}{\sqrt{3}}x$ and enlargement with centre O and scale factor 2

2 a) rotation about O through 45° clockwise and enlargement with centre O and scale factor 2√2; rotation about O through 45° anticlockwise and enlargement with centre O and scale factor 3√2

2 b) both equal to enlargement with centre O and scale factor 12

3 a) stretch parallel to x-axis with scale factor 2; stretch parallel to y-axis with scale factor 5

3 b) both equal to a two-way stretch with scale factor 2 parallel to x-axis and scale factor 5 parallel to y-axis

4 a) √2; 45° clockwise   **b)** enlargement with scale factor 2, centre the origin and rotation through 90° clockwise

4 c) enlargement with scale factor 4 and rotation through 180°; enlargement with scale factor 8 and rotation through 90° anticlockwise; enlargement with scale factor 16, all centre the origin

5 a) rotation about O through 60° anticlockwise   **b)** 6   **6 a)** $\begin{pmatrix} 0 & 1 \\ -1 & 0 \end{pmatrix}$ **b)** rotation about O through 90° clockwise

7 a) i) $\begin{pmatrix} -\frac{1}{2} & \frac{\sqrt{3}}{2} \\ -\frac{\sqrt{3}}{2} & -\frac{1}{2} \end{pmatrix}$ **ii)** $\begin{pmatrix} 1 & 0 \\ 0 & 1 \end{pmatrix}$ **b)** rotation about O through 120° anticlockwise

8 a) (3, −4), (4, 3)   **b)** rotation about O through 53° (approx) clockwise   **9 a)** rotation about O through 30° anticlockwise

9 b) $\begin{pmatrix} -\frac{1}{2} & \frac{\sqrt{3}}{2} \\ \frac{\sqrt{3}}{2} & \frac{1}{2} \end{pmatrix}$ **c) i)** $\begin{pmatrix} -\frac{\sqrt{3}}{2} & \frac{1}{2} \\ \frac{1}{2} & \frac{\sqrt{3}}{2} \end{pmatrix}$ **ii)** reflection in $y = (\tan 75°)x$

10 a) reflection in $y = x$; reflection in x-axis   **b)** rotations about O through 90° clockwise and anticlockwise   **c)** identity transformation

Chapter 5

Check in

1 c), a), b), a), b)   **2 a)** 7 **b)** 0 **c)** $-\frac{3}{2}$ **d)** $\frac{2}{3}$ **3 a)** $x = -\frac{3}{4}$ **b)** $x > -\frac{3}{4}$ **c)** $x = \frac{1}{2}$ or 1 **d)** $\frac{1}{2} < x < 1$ **e)** $x < \frac{1}{2}, x > 1$

Note that the answers for Exercise 5A to Exercise 5E contain graphs, and are not included here. You should check your answers on a graphics calculator.

Exercise 5F

1 $(-\frac{1}{2}, \frac{5}{13})$ **2** $(1, \frac{1}{2}), (5, \frac{5}{6})$ **3** $(1, -1), (-2, \frac{1}{2})$ **4** $(0, 1), (-\frac{6}{5}, -\frac{1}{23})$ **5** $(1, 1), (\frac{5}{3}, -3)$

Exercise 5G

1 a) $x > -1, x < -2$ **b)** $x > 3$ **c)** $-10 < x < -3$ **d)** $x > 5$ and $x < -14$ **e)** $-0.5 < x < -0.3$ **f)** $\frac{1}{5} < x < \frac{6}{11}$

2 a) $x < -2, -1 < x < 0$ **b)** $x > 8, 2 < x < 3$ **c)** $5 < x < \frac{1}{2}(3 + \sqrt{65}), \frac{1}{2}(3 - \sqrt{65}) < x < -1$

2 d) $x < -7, 3 < x < \frac{44}{13}$ **e)** $-2 < x < -1\frac{1}{2}$ **3** $1 > x > -2$ **b)** $x > 1, x < -\frac{3}{2}$ **c)** $x < -2$ $(x \neq -1)$

Check out

1 a) $x = -\frac{3}{2}$ **b)** $y = \frac{3}{2}; x = -\frac{3}{2}, x = 1, y = 0$ **c)** $x = -\frac{3}{2}, y = 0$ **2 a)** $(\frac{2}{3}, 0), (0, -\frac{2}{3})$ **b)** $(\frac{2}{3}, 0), (1, 0), (0, \frac{2}{3})$ **c)** $(0, \frac{2}{3})$

4 $(-2, 0), (\frac{1}{2}, 5)$ **5** $-\frac{1}{2} < x < 0, 1 < x < 2$

Revision exercise 5

1 Asymptotes $x = 2, y = 1$ **2 a)** $y = 1$ **b) ii)** $(0, 0), (-2, 4)$ **3 a)** $y = 2 - \dfrac{3}{x + 2}$ **b)** $x = -2, y = 2; (0, \frac{1}{2}), (-\frac{1}{2}, 0)$

4 a) $x = -1, y = 2, (\frac{1}{2}, 0), (0, -1)$ **b)** $x < -3$ or $x > -2$ **5** 4, 1   **6** $x < 2$ or $x > 8$ **7 b)** $(-3, -\frac{1}{12})$ **c)** $x < 2, x > \frac{9}{2}$

8 b) $\frac{3}{4} < x < 1$ and $x > 2$ **9 b)** $y = 1$ **e)** $x < -1, x > 1$

Chapter 6

Check in

1 a) real distinct   b) equal   c) none   d) none   e) equal   f) real distinct   **2** a) $-1, -3$ b) $-2, -2$ c) $1, \frac{1}{2}$ d) $1, \frac{1}{3}$

3 a) $y = (x - 2)^2$ b) $y = x^2 - 3$ c) $y = \frac{1}{4}x^2$ d) $y = 4x^2$ e) $x = y^2$

Exercise 6A

1 a) $(0, 0), (\frac{4}{5}, 4)$ b) $(20, 20)$ **2** a) $(0, 0), (2, 8)$ b) $(72, 48)$ c) No points   **3** $(0, 0), (\frac{12}{5}, 12)$ b) $(0, 0), (\frac{5}{12}, 5)$

3 c) No points   d) $(15, 30)$

Exercise 6B

6 $x^2 = 8y$ **7** $x^2 = 1764y$

Exercise 6C

1 a) $(\pm\frac{10}{\sqrt{2029}}, \pm\frac{90}{\sqrt{2029}})$ b) $(0, 2), (\frac{100}{29}, -\frac{42}{29})$ **2** a) $(\pm\frac{4}{\sqrt{65}}, \pm\frac{16}{\sqrt{65}})$ b) $(0, 2)$ c) No roots

3 a) $(\pm\frac{5}{\sqrt{26}}, \pm\frac{45}{\sqrt{26}})$ b) $(-4.890, 1.877), (4.981, 0.780)$ **4** $\frac{y^2}{25} + \frac{x^2}{4} = 1$ **5** $\frac{(x-3)^2}{9} + \frac{(y+5)^2}{4} = 1$

6 $\frac{x^2}{144} + \frac{y^2}{25} = 1$ **7** $\frac{x^2}{25} + \frac{y^2}{64} = 1$ **8** $\frac{x^2}{225} + \frac{y^2}{4} = 1$ **9** $\frac{x^2}{900} + \frac{y^2}{196} = 1$

Exercise 6D

1 a) $(\pm\frac{90}{\sqrt{299}}, \pm\frac{10}{\sqrt{299}})$ b) $(9.00, 3.00), (19.6, -7.6)$ **2** a) $(\pm\frac{12}{\sqrt{35}}, \pm\frac{3}{\sqrt{35}})$ b) $(2, 0), (-\frac{74}{35}, \frac{36}{35})$

3 a) $(\pm\frac{100}{\sqrt{609}}, \pm\frac{20}{\sqrt{609}})$ b) $(\frac{401}{10}, -\frac{399}{8})$ **4** $\frac{y^2}{16} - \frac{x^2}{4} = 1$ **5** $\frac{(x-4)^2}{16} - \frac{(y-3)^2}{81} = 1$ **6** $\frac{x^2}{144} - \frac{y^2}{49} = 1$

7 $\frac{x^2}{16} - \frac{y^2}{1875} = 1$ **8** $\frac{y^2}{128} - \frac{x^2}{25} = 1$ **9** $\frac{y^2}{32} - \frac{x^2}{441} = 1$

Exercise 6E

1 a) $(2, 18), (-2, -18)$ b) $(6, 6)$ **2** a) $(\frac{5}{2}, 10), (-\frac{5}{2}, -10)$ b) $(10, \frac{5}{2})$ Line is a tangent.

3 a) $(\frac{2}{3}, 6), (-\frac{2}{3}, -6)$ b) $(6, \frac{2}{3})$ Line is a tangent   **4** $xy = 36$ **5** $(x - 4)(y - 3) = 49$ **6** $xy = 324$ **7** $xy = 405$

8 $xy = 243$ **9** $xy = 1620$

Check out

1 a) Hyperbola   b) ellipse   c) parabola   d) rectangular hyperbola   **3** a) $(1, 9), (\frac{9}{4}, 4)$ b) $(\frac{3}{2}, 6)$ c) none

4 a) $(y - 1)^2 = 24(x + 2)$ b) $xy = 2400$ c) $\frac{y^2}{100} - \frac{x^2}{25} = 1$

Revision exercise 6

1 a) rectangular hyperbola

1 b) $(4, 4), (-4, -4)$ $y = x + 2$ has 2 points of intersection; $y = x + 4$ is a tangent; $y = x + 6$ has no points of intersection

3 b) i) $\frac{x^2}{9} + y^2 = 1$ ii) $\frac{x^2}{9} + \frac{y^2}{16} = 1$ iii) $\frac{(x-1)^2}{9} + \frac{(y-2)^2}{16} = 1$

4 b) i) $(x - 1)^2 - (y - 2)^2 = 1$ ii) $(\frac{1}{3}x - 1)^2 - (y - 2)^2 = 1$ iii) $(\frac{1}{3}x - 1)^2 - (\frac{1}{4}y - 2)^2 = 1$

5 a) $(y - 1)^2 - x^2 = 1$ b) $y = 0, y = 2$ c) $k < 0, k > 2$ **6** b) $x^2 = 8y$ c) Translation 2 units in positive x direction

7 a) Intercepts $(0, \pm7), (\pm4, 0)$ b) Stretch parallel to x-axis with scale factor $\frac{1}{2}$, translation 3 units in positive y direction

8 6   **9** a) G_1

Chapter 7

Check in

1 $x = 1, y = -2$ **2** $x = \dfrac{3 \pm \sqrt{41}}{4}$

Exercise 7A

1 a) $-i$ b) 1 c) -1 d) i **2** a) $10 - 2i$ b) $1 - i$ c) $-1 + 2i$ d) $2 + 55i$ e) $1 + i$ f) $3 + 8i$ g) $10 + 18i$

2 h) $18 + 13i$ **3** a) $3 + 11i$ b) $26 + 2i$ c) $74 + 7i$ d) $42 - 24i$ e) $10 + 11i$ f) $11 - 29i$

Exercise 7B

1 a) $-1 \pm \sqrt{3}\,i$ b) $\frac{3}{2} \pm \frac{\sqrt{15}}{2}i$ c) $-\frac{1}{4} \pm \frac{\sqrt{7}}{4}i$ d) $1 \pm \frac{\sqrt{2}}{2}i$ **2** a) $-2 \pm \sqrt{3}\,i$ b) $-1 \pm \sqrt{5}\,i$ c) $-\frac{3}{2} \pm \frac{3}{2}i$ d) $\frac{5}{2} \pm \frac{5\sqrt{3}}{2}i$

3 $x^2 + 14x + 60 = 0$ **4** $x^2 - 12x + 81 = 0$ **5** $4x^2 + 59x + 289 = 0$ **6** $9x^2 + 41x + 255 = 0$

7 a) $x^2 + 3x + 14 = 0$ b) $18x^2 + 9x + 7 = 0$ c) $4x^2 + 19x + 49 = 0$ d) $x^2 - 11x + 42 = 0$

8 a) $3x^2 + 8x + 128 = 0$ b) $3x^2 + x + 2 = 0$ c) $9x^2 + 44x + 64 = 0$ d) $3x^2 - 14x + 208 = 0$

9 $x^2 + 10x + 75 = 0$ **10** $8x^2 - 5x + 1 = 0$

Exercise 7C

1 a) $2 - 9i$ b) $-1 - 6i$ **2** a) $-3 + \frac{1}{3}i$ b) $\frac{5}{7} - \frac{2}{7}i$ **3** a) $2 + i$ b) $-\frac{7}{2} + \frac{7}{2}i$ c) $\frac{16}{13} - \frac{22}{13}i$ **4** $\frac{3}{4} - \frac{1}{4}i$

Check out

1 a) $8 - 2i$ b) $5 - 3i$ c) $7 + i$ **2** a) 6 b) $8i$ c) 25   **3** a) $x = \pm 3i$ b) $x = -2 \pm i$ c) $x = 3 \pm i\sqrt{7}$ **4** $1, -2$

Revision exercise 7

1 b) $-4, -8i, 16$ **2** a) $2x + 2 = y, x - 4 = -3$ (or $x = 1$) b) $x = 1, y = 4$ **3** a) $x - y = 1, x + y = 3$ b) $x = 2, y = 1$

4 a) $3x - iy$ b) $3 - 2i$ **5** a) $-x + 7iy$ b) $-5 + 3i$ **7** a) i) $8 + 6i$ ii) $2 + 14i$ **8** a) i) $-5 + 12i$ ii) $-119 - 120i$

8 b) i) 39 ii) $2 - 3i$ or -1 or -3 **9** a) i) -12 ii) $-\frac{12}{169}$ b) $169x^2 + 12x + 13 = 0$ c) $-2 \pm 3i$

10 $z = k + (2 - k)i$ for all real k

Chapter 8

Check in

1 a) 2 b) 1 c) 0   **2** a) 0 b) 0 c) 2 d) none   **3** a) i) $\dfrac{1}{x}$ ii) $\dfrac{1}{x^2}$ iii) $\dfrac{1}{\sqrt{x}}$ iv) $\dfrac{1}{x\sqrt{x}}$ b) i) $x^{-\frac{1}{2}}$ ii) $x^{-\frac{5}{2}}$ iii) $x^{\frac{1}{2}}$

4 a) $-x^{-1} (+c)$ b) $2x^{\frac{1}{2}} (+c)$ c) $-2x^{-\frac{1}{2}} (+c)$ d) $-\frac{2}{3}x^{-\frac{3}{2}} (+c)$

Exercise 8A

1 $-3 + h; -3$ **2** $7 + h; 7$ **3** $-25 + 4h; y = {}^-25x - 33$ **4** $23 + 2h; y = 23x - 58$ **5** $6a + 7$ **6** $5 - 8a$

Exercise 8B

1 $49 + 18h + 2h^2, 49$ **2** $4 + 6h + 4h^2 + h^3, 4$ **3** $4a^3 - 6a$ **4** $80 + 80h + 40h^{\frac{1}{2}} + 10h^3 + h^4, y = 80x - 128$

5 $-\dfrac{1}{4(4 + h)}; 16y + x = 8$

Exercise 8C

1 $\frac{3}{2}$ **2** Does not exist   **3** Does not exist   **4** Does not exist   **5** Does not exist   **6** $\frac{3}{2}$

Check out

1 55   **2** a) and c) are improper integrals   **3** 2, not possible, not possible, $\frac{2}{3}$

Revision exercise 8

1 b) 8   **2** b) 26   **3** a) 3 c) -8 **4** a) $3h + h^2$ c) $3, 6$ **5** a) $3h + h^2$ c) $3, 6$ **6** a) $\frac{3}{4}$ b) not possible

7 a) $\frac{3}{2}$ b) not possible   **8** a) not possible b) 3   **9** a) -6 b) $\frac{40}{27}$ **10** d) One of the limits of integration is $-\infty$

FP1

Chapter 9

Check in

1 a) $\dfrac{\pi}{4}$ b) 2π c) $\dfrac{2\pi}{3}$ d) $\dfrac{\pi}{6}$ **2** a) $\dfrac{\sqrt{3}}{2}$ b) $\dfrac{1}{\sqrt{2}}$ c) $\dfrac{\sqrt{3}}{2}$

Exercise 9A

1 a) $360n° \pm 30°$ b) $2n\pi \pm \dfrac{\pi}{6}$ **2** a) $360n° \pm 120°$ b) $2n\pi \pm \dfrac{2\pi}{3}$ **3** a) $180n° \pm 22.5°$ b) $n\pi \pm \dfrac{\pi}{8}$

4 a) $120n° \pm 45°$ b) $\dfrac{2}{3}n\pi \pm \dfrac{\pi}{4}$ **5** a) $72n°$ b) $\dfrac{2n\pi}{5}$ **6** a) $360n° \pm 60°$; $360n° \pm 120°$ b) $2n\pi \pm \dfrac{\pi}{3}$, $2n\pi \pm \dfrac{2\pi}{3}$

7 $\dfrac{2}{3}n\pi + \dfrac{2\pi}{9}$, $\dfrac{2n\pi}{3}$ **8** $n\pi + \dfrac{\pi}{8} \pm \dfrac{\pi}{12}$ **9** $n\dfrac{\pi}{2} + \dfrac{\pi}{24} \pm \dfrac{\pi}{16}$ **10** $n\pi + 0.5 \pm 0.886$

FP1

Exercise 9B

1 a) $180n° + (-1)^n 60°$ b) $n\pi + (-1)^n\dfrac{\pi}{3}$ **2** a) $180n° + (-1)^n 45°$ b) $n\pi + (-1)^n\dfrac{\pi}{4}$

3 a) $90n° + (-1)^{n+1}15°$ b) $\dfrac{n\pi}{2} + (-1)^{n+1}\dfrac{\pi}{12}$ **4** $60n° + (-1)^n 15°$ b) $\dfrac{n\pi}{3} + (-1)^n\dfrac{\pi}{12}$

5 a) $72n° - 18°$ b) $\dfrac{2n\pi}{5} - \dfrac{\pi}{10}$ **6** $180n° + (-1)^n 60°$; $180n° + (-1)^n 120°$ b) $n\pi + (-1)^n\dfrac{\pi}{3}$; $n\pi + (-1)^n\dfrac{2\pi}{4}$

7 $\dfrac{n\pi}{3} + \dfrac{\pi}{9} + (-1)^n\dfrac{\pi}{18}$ **8** $\dfrac{n\pi}{2} + \dfrac{\pi}{8} + (-1)^n\dfrac{\pi}{6}$ **9** $\dfrac{n\pi}{2} + 0.5 + (-1)^{n+1}0.206$ **10** $2n\pi - \dfrac{\pi}{2}$, $n\pi + (-1)^{n+1}\dfrac{\pi}{6}$

11 $2n\pi + \dfrac{\pi}{2}$, $n\pi + (-1)^n 0.253$ **12** $n\pi + (-1)^{n+1}\dfrac{\pi}{6}$, $n\pi + (-1)^n 0.340$ **13** $\dfrac{2n\pi}{3} + \dfrac{\pi}{3}$

Exercise 9C

1 a) $180n° + 60°$ b) $n\pi + \dfrac{\pi}{3}$ **2** a) $180n° - 45°$ b) $n\pi - \dfrac{\pi}{4}$ **3** a) $30n° + 7.5°$ b) $\dfrac{n\pi}{6} + \dfrac{\pi}{24}$

4 $60n° - 10°$ b) $\dfrac{n\pi}{3} - \dfrac{\pi}{18}$ **5** $36n° + 9°$ b) $\dfrac{n\pi}{5} + \dfrac{\pi}{20}$ **6** $180n° \pm 30°$ b) $n\pi \pm \dfrac{\pi}{6}$ **7** $\dfrac{n\pi}{3} + \dfrac{7\pi}{36}$

8 $\dfrac{n\pi}{2} + \dfrac{7\pi}{24}$ **9** $\dfrac{n\pi}{4} - \dfrac{\pi}{12}$ **10** $\dfrac{n\pi}{4} + \dfrac{\pi}{16}$ **11** $n\pi + \dfrac{\pi}{12}$ **12** $\dfrac{n\pi}{3} + \dfrac{5\pi}{36}$ **13** $\dfrac{n\pi}{4}$

Check out

1 $x = 70° + 180n°$ or $-50° + 180n°$ **2** $x = \dfrac{2n\pi}{3}$ or $\dfrac{2n\pi}{3} + \dfrac{\pi}{9}$ or $\dfrac{n\pi}{3} + \dfrac{\pi}{9} + (-1)^n\dfrac{\pi}{9}$ **3** $x = 5° + 18n°$

Revision exercise 9

1 a) $180°$; $(90 + 180n)°$; $180n°$ b) $n\pi$; $\dfrac{\pi}{2} + n\pi$; $n\pi$ **2** a) $(30 + 360n)°$, $(150 + 360n)°$; $(\pm 60 + 360n)°$; $(45 + 180n)°$

2 b) $\dfrac{\pi}{6} + 2n\pi$, $\dfrac{5\pi}{6} + 2n\pi$, $\pm\dfrac{\pi}{3} + 2n\pi$, $\dfrac{\pi}{4} + n\pi$ **3** a) $(30 + 180n)°$, $(60 + 180n)°$; $(\pm 15 + 180n)°$; $(30 + 90n)°$

b) $\dfrac{\pi}{6} + n\pi$, $\dfrac{\pi}{3} + n\pi$, $\pm\dfrac{\pi}{12} + n\pi$, $\dfrac{\pi}{6} + \dfrac{n\pi}{2}$ **4** $-\dfrac{5\pi}{12} + 2n\pi$, $-\dfrac{13\pi}{12} + 2n\pi$, $\dfrac{11\pi}{12} + 2n\pi$, $-\dfrac{5\pi}{12} + 2n\pi$, $n\pi$ **5** $-\dfrac{\pi}{3} + n\pi$

6 a) $-\dfrac{1}{\sqrt{3}}$ b) $-\dfrac{\pi}{6} + n\pi$ **7** $(40.5 + 120n)°$, $(92.8 + 120n)°$ **8** $(57 + 60n)°$ **9** $\dfrac{\pi}{2} + 2n\pi$, $-\dfrac{5\pi}{6} + 2n\pi$ **10** b) $\dfrac{\pi}{3} + n\pi$

Chapter 10

Check in

1 6 cm   **2** a) $3x^2$ b) $10x + 2$ c) $28x^3$

Exercise 10A

1 $(2.25, 2.5)$ **2** $(0.5, 0.75)$ **3** $(0.75, 1)$ **4** $(3.25, 3.5)$ **5** $(0.5, 0.625)$ **6** $(0.55, 0.6)$

Exercise 10B
(All answers given to 5 decimal places)

1 1.08333   **2** 0.33333   **3** 1.43224   **4** 2.58834   **5** 0.41207   **6** 3.13844

Exercise 10C

1 2.22   **2** 2.52   **3** 1.95 **4** -1.13 **5** 1.41

Exercise 10D
(All answers given to 4 significant figures)

1 2.104   **2** 1.001   **3** 3.128   **4** 3.644   **5** 2.017   **6** 4.028   **7** 1.011

Check out

1 $f(1.1) \approx -0.0088 < 0$, $f(1.2) \approx 0.13 > 0$ **2** 1.5   **3** 0.722   **4** 1.490   **5** 3.062

Revision exercise 10

3 b) ii) $1.3 - 1.35$ iii) 1.3   **4** c) 0.149 d) 0.86   **5** a) 2.81 b) ii) $2.2 - 2.3$ **6** a) 1.33 b) 1.6   **7** 0.71   **8** 3.35
9 5.98   **10** 1.0041   **11** 1.31

Chapter 11

Check in

2 $-\frac{1}{2}, 3.5$ **3** $y = -\frac{1}{2}x + \frac{7}{2}$ **4** a) 1.5 b) 7   **5** a) $\log_{10} a + \log_{10} b$ b) $n\log_{10} a$

Exercise 11A

1 $4, 5; x = 4t^2 + 5$ **2** 10   **3** $2; 7$ **4** $2; 3$ **5** 2   **6** $3; 7$ **7** $6; 2$

Exercise 11B

1 a) ii) $a = 20, b = 1.2$ b) 20 grams

1 c) 1 126 950 grams, i.e. just over 1 tonne. The bacteria are likely to stop multiplying exponentially before 60 days have passed. For example, they may run out of food.

2 a) ii) $a = 30, b = 0.9$ b) 24.3 grams c) 14.3 grams   **3** $a = 7, b = 1.5$ **4** $a = 5.1, b = 1.8$

5 a) 200 grams; b) mass when $t = 3$ should have read 26.5 grams.

Check out

1 e.g. a) $X = \dfrac{1}{x}$, $Y = \dfrac{1}{y}$; b) $X = x^3$, $Y = y^2$ c) $X = x^2$, $Y = \dfrac{y^2}{x}$
2 a) $\log y = n \log x$ b) $\log y = (\log k)x$ c) $\log u = \log k - n \log v$

Revision exercise 11

1 a)

X	4	9	16	25
Y	3	7	13	20

d) 0.8 e) 0.8   **2** a)

X	1.44	4.41	5.76	9.61	12.96
Y	14.44	11.56	10.24	6.25	3.24

2 d) 16, 4   **3** a) $\log_{10} T = \log_{10} a + b \log_{10} L$ d) i) 4.7 ii) 3.9, 0.52   **4** $7.39x^{1.33}$

FP1

5 a) $\log_{10} Q = \log_{10} a + b \log_{10} x$ b) i) $\log_{10} x$: $-0.40, -0.30, -0.22 -0.15, -0.10$ ii) $\log_{10} Q$: $0.24, 0.48, 0.68, 0.84, 0.99$

5 c) i) 3.6 ii) 16, 2.5   **6** a) 331 b) i) $\log_{10} N = \log_{10} a + (\log_{10} b)t$

6 b) ii) 251, 1.06 Bacteria will start to die as well as reproduce; will reach saturation

7 a) $\log_{10} P = \log_{10} k + n \log_{10} I$ d) i) 2.0 ii) 21

8 a) i) 5.;3, 5.6, 5.8, 6.1, 6.7, 7.4 b) Graph of $\dfrac{x}{t}$ against t is a straight line c) 4.4, 0.58

9 a) $\log_{10} y = \log_{10} a + n \log_{10} x$ c) The points lie on a straight line d) i) 4.1 ii) 2.4

Practice paper

1 a) i) $\begin{pmatrix} 12 & 0 \\ 0 & 12 \end{pmatrix}$ ii) $\begin{pmatrix} 12 & 24 \\ 36 & 48 \end{pmatrix}$ b) Enlargement with centre O and scale factor 12   **2** a) $x - i(y + 1)$ b) $-i$

3 a) $\frac{1}{2}, -1$ b) i) $2\frac{1}{4}$ ii) $4\frac{1}{4}$ **4** a) $4h + h^2$ b) 4   **6** 3.20   **7** a) $\dfrac{1}{2}, \dfrac{\sqrt{3}}{2}, \dfrac{1}{\sqrt{3}}$ b) $\dfrac{\pi}{2} + 2n\pi, \dfrac{\pi}{6} + 2n\pi$

8 a) $y = 1, x = -1, x = 2$ b) $(0, -\frac{3}{2}), (1, 0), (3, 0)$ c) $(\frac{5}{3}, 1)$ **9** b) Stretch parallel to x-axis with scale factor 2 d) Tangent

FP1

Appendix: An alternative approach to complex numbers

Some students of module FP1 may be happy to accept the notion that there is a number called i, outside the system of real numbers, which has the property that $i^2 = -1$, and which can be combined with real numbers to form the system of complex numbers.

Others may be dissatisfied by the idea of working with a number which is purely imaginary, and may be glad to know that 'complex numbers' are just as real as 'real numbers'.

If you intend to study module FP2 you may find the ideas in that module easier to understand if you think of complex numbers as numbers which represent certain transformations.

Rotations

Consider a familiar situation: a knob on an electric cooker which can be clicked into any one of four positions by rotations through multiples of 90°.

One click anticlockwise can be identified with the number i.

Two clicks are equivalent to a rotation through 180°, for which the matrix is $\begin{bmatrix} -1 & 0 \\ 0 & -1 \end{bmatrix}$.

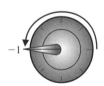

This is -1 times the identity matrix **I**, leading to the relationship

$$i^2 = -1.$$

Three clicks anticlockwise will have the same effect as one click clockwise, which you can identify with the complex number $-i$.

Performing this operation four times returns the knob to its original position, corresponding to the identity transformation with matrix **I**.

Thus $i^1 = i$, $i^2 = -1$, $i^3 = -i$, $i^4 = 1$, $i^5 = i$, and so on.

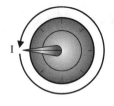

The four numbers i, -1, $-i$ and 1 form a **closed** system in which multiplying any two of the numbers always gives an answer within the system. Closure is one of the essential requirements for a successful number system.

This small system can be extended to include **all** rotations about the origin. A rotation through an angle α followed by a rotation through an angle β will result in a rotation through the angle $\alpha\beta$.

The matrix version of this statement is:

$$\begin{bmatrix} \cos \beta & -\sin \beta \\ \sin \beta & \cos \beta \end{bmatrix} \begin{bmatrix} \cos \alpha & -\sin \alpha \\ \sin \alpha & \cos \alpha \end{bmatrix} = \begin{bmatrix} \cos(\alpha + \beta) & -\sin(\alpha + \beta) \\ \sin(\alpha + \beta) & \cos(\alpha + \beta) \end{bmatrix}$$

It will still be true that combining any two members of the system gives a result that is also a member of the system. Hence, if you represent the rotations by matrices, you have a system that is closed for multiplication.

This system has other desirable features from the point of view of the mathematician. For example, the multiplication is **commutative**, so reversing the order of the two rotations has no effect on the result. It is also possible to **divide** these matrices, since a rotation through an angle α **clockwise** will reverse the effect of a rotation through α anticlockwise.

However, **adding** the matrices in this system will not give a matrix belonging to the system. For example,

$$\begin{bmatrix} 1 & 0 \\ 0 & 1 \end{bmatrix} + \begin{bmatrix} 1 & 0 \\ 0 & 1 \end{bmatrix} = \begin{bmatrix} 2 & 0 \\ 0 & 2 \end{bmatrix}$$

which is not the matrix of a rotation. Therefore the system is not closed for addition.

A system of numbers in which you can multiply and divide, but not add and subtract, is of very limited value. To obtain a really satisfactory system you need to combine the rotations with enlargements.

Rotations-and-enlargements

If you combine a rotation through an angle α with an enlargement of scale factor r, you obtain a transformation with matrix

$$\begin{bmatrix} r \cos \alpha & -r \sin \alpha \\ r \sin \alpha & r \cos \alpha \end{bmatrix}.$$

This matrix is of the form $\begin{bmatrix} a & -b \\ b & a \end{bmatrix}$, and indeed **any** matrix of this form can be expressed in the form $\begin{bmatrix} r\cos\alpha & -r\sin\alpha \\ r\sin\alpha & r\cos\alpha \end{bmatrix}$, with $r = \sqrt{a^2 + b^2}$

and with the angle α chosen so that $\cos\alpha = \dfrac{a}{r}$ and $\sin\alpha = \dfrac{b}{r}$.

You can add any two matrices of the form $\begin{bmatrix} a & -b \\ b & a \end{bmatrix}$ and the result will

be a matrix of the same form:

$$\begin{bmatrix} a & -b \\ b & a \end{bmatrix} + \begin{bmatrix} c & -d \\ d & c \end{bmatrix} = \begin{bmatrix} a+c & -(b+d) \\ b+d & a+c \end{bmatrix}.$$

And if you multiply any two matrices of this form the result will again
be a matrix of the same form:

$$\begin{bmatrix} a & -b \\ b & a \end{bmatrix}\begin{bmatrix} c & -d \\ d & c \end{bmatrix} = \begin{bmatrix} ac-bd & -(ad+bc) \\ ad+bc & ac-bd \end{bmatrix}.$$

If one matrix represents a rotation through an angle α with an
enlargement of scale factor r, and the other represents a rotation
through an angle β with an enlargement of scale factor s, the result
will be a rotation through angle $(\alpha + \beta)$ with an enlargement of scale
factor rs.

> This result is not surprising if you think of combining two rotations-and-enlargements.   **FP1**

In other words, you **add** the angles and **multiply** the scale factors.
Using the terminology of complex numbers, you **add** the **arguments**
and **multiply** the **moduli**.

Powers and roots

From the preceding argument you can deduce:

✦ how to divide any complex number by any non-zero complex
  number (subtract the angles and divide the scale factors);
✦ how to square any complex number (double the angle and square
  the scale factor);
✦ how to find the two square roots of any non-zero complex number
  (for one square root, halve the angle and take the square root of
  the scale factor; for the other, add π radians to the angle previously
  found and keep the scale factor unchanged from the first square
  root).

The technique just described can be extended to finding the three
cube roots, the four fourth roots, and so on, of any non-zero complex
number.

> Every quadratic equation has two roots, every cubic equation has three roots, and so on.

Possible snags

When mathematicians first developed complex numbers they had to
check whether **all** the techniques used in handling real numbers would
still be valid.

Using the approach outlined here, you can establish that complex numbers obey all the rules of algebra, such as

$$x(y + z) = xy + xz \quad \text{and} \quad (xy)^2 = x^2y^2.$$

Hence complex numbers form a system which behaves very much like the system of real numbers, with the enrichment that every number in the system has square roots, cube roots, and so on.

There are bound to be losses associated with this considerable gain.

✦ The complex numbers cannot be **ordered**, so that the statement $a < b$ has no meaning if a and b are complex numbers.

✦ Be careful with the use of the notation $\sqrt{x}$ where x is a complex number. It is **not** possible to define this in such a way that the rule $\sqrt{xy} = \sqrt{x}\,\sqrt{y}$ will remain true as it does for positive real numbers. The remedy is to put the symbol $\pm$ before the square root symbol, since the two square roots are always opposites of each other.

Try it with $x = y = -1$.

FP1

Index